The Sensitive Scientist

Report of a British Association Study Group

DAVID MORLEY

SCM PRESS LTD

334 01386 0

First published 1978
by SCM Press Ltd
56 Bloomsbury Street, London
Phototypeset by Filmtype Services Ltd, Scarborough
Printed in Great Britain by Fletcher & Son Ltd, Norwich

Contents

Foreword

Some time in 1973 Dr Magnus Pyke and Sir Lincoln Ralphs on behalf of the British Association for the Advancement of Science approached the British Council of Churches and invited them to discuss the possibility of a study group on the general theme of science and ethics. As a result of these initial talks, the British Association resolved to set up a working group. The membership of the study group was determined at that consultation, and by subsequent co-option. In particular, we decided that although the initial approach had been made to the British Council of Churches, the group should be widely representative alike of different scientific concerns and of different approaches to ethics, Christian and non-Christian.

It is important to see what we are not professing to do.

We are not trying to discuss all or even the most important ethical questions which confront scientists. We have been selective, and have in general kept close to areas where one or more of us had trodden already.

We are not seeking to repeat what others have said or written elsewhere about science and ethics. We are well aware of the growing corpus of literature in this area, but have not felt it necessary to provide a conspectus or survey of this. The references at the end of each chapter are not a bibliography to the theme of that chapter but references in the strict sense.

We do not profess to solve the problems we discuss. We want rather (a) to show that there are ethical problems to which some practising scientists are sensitive and some are sometimes blind; (b) to reveal something of the complexity of these problems, and the seemingly irreconcilable clashes of value they involve; (c) to suggest ways in which the issues may be clarified, and the problems faced constructively.

We are aware that there is much more which we might have done; with our inevitably limited resources this is what we feel able to offer. It is a contribution by a miscellaneous group of concerned people to a debate which is being and will go on being conducted at other levels.

Our method of procedure has been to invite individual members to submit discussion papers out of their own concerns and experience. These have been exposed to general discussion, and presented afresh in the light of that discussion. We also read, and occasionally invited, papers from outside the committee. Out of these discussions a pattern began to emerge, and we could discern the skeleton of the final report. Dr Morley then began to draft individual chapters, and each of these was submitted to the same process of discussion and re-presentation. Although individual members of the committee might disagree with the emphasis at some points, the document in its final form has the endorsement of the whole committee.

We should like to say how much we ourselves have gained from meeting together. We owe a considerable debt to the British Association for bringing us together and looking after us, and in particular to Dr David Morley, who has drafted the final report (which has sometimes involved extensive writing and rewriting), for his fairness in recording, his own constructive contributions, his extremely hard work, and his witty and incisive style of presentation.

Christmas 1976   JOHN FERGUSON
                                              Chairman

Members of the British Association Study Group on Science and Ethics

Acknowledgments

The Committee would like to thank all those individuals and organizations who gave written or verbal evidence, including FRAME, Canon D. E. Jenkins (who was forced to resign from the Committee after the first meeting due to pressure of other duties), Dr E. S. Johnson, Sir Peter Medawar, the Medical Research Council, Dr G. E. Paget, Mr B. A. Paskins, Mr John Rivers, the Royal College of Physicians, and Mrs Frances D'Souza. Thanks should also go to Mrs Shirley Parker who typed not only the final report but also the previous drafts and many of the supporting committee papers.

Chapter 4 is based on the article, 'Ethics of Selective Treatment of Spina Bifida' which originally appeared in *The Lancet*, 1, 1975.

1 Introduction

Science, in the widest sense of the word, covers a broad spectrum of activities. At one end of that spectrum lies the search for pure knowledge; at the other lies the application of that knowledge to social problems. Whilst there is no sharp dividing line between these two aspects of science, they are nevertheless sufficiently distinct that one can distinguish between pure science (often, rather confusingly, referred to simply as 'science') and applied science or technology.

This distinction is fundamental to the attitude of many scientists to ethics. They believe, quite rightly, that whenever science is applied to social problems then ethical judgments are inevitably involved. But they also believe that pure science is not concerned with values, but only with facts. Scientific knowledge, based on observation and theory, and on the assessment of what is true and what is probable, is regarded as being totally objective: not merely complementing subjective, value-ridden judgments but even replacing them altogether. This is in complete contrast to ethics. For the latter, value judgments are absolutely essential; they are the stuff of any system of morals and human conduct.

It is for this reason, perhaps, that many scientists genuinely believe that pure science, for which value judgments are an anathema, has nothing to do with ethics. They fail to see that science itself is grounded in value judgments, which they either take for granted or, if they become aware of them, take as self-evident. They assume that 'fact' and 'value' are wholly distinct concepts. This, as will be discussed later, can be very misleading.

Science is a human activity. It is not just the collection and analysis of data, nor the construction of theories. It also involves people and their values. The pure scientist does not exist in isolation from the rest of society: he spends its money, educates its children, heals its sick and feeds its hungry, develops its consumer products – indeed, there is almost no aspect of modern life which is untouched by science and the scientist. He is also a human being and a citizen, who makes mistakes, collaborates and competes with his scientific colleagues, tries to persuade others to do what he thinks is right;

who, in short, displays all the qualities and failings of the human race. Man is an ethical animal, and so is the scientist.

When, late in 1973, the British Council of Churches and the British Association first began to explore the possibility of conducting a study on science and ethics, these thoughts were uppermost in their minds. Almost every aspect of science and technology touches on ethics in some way, and yet, despite the growing interest in the social responsibilities of scientists, many people still failed to realize how basic, how essential ethical discussion was to the business of science. The study was conceived as an attempt to clarify these questions, and to show, by considering the particular case studies and then discussing the deeper issues which underlie them, how intimately ethical judgments are involved in science. It is the results of this study which form the material for this report.

This is not a book on ethical theory, but an attempt to show how ethics permeates those real and practical issues which arise from science. Nor is it a book of answers. The study of ethics may clarify problems, and even rule out certain answers, but it rarely solves problems completely. One is almost always left with a difficult choice. We live in a pluralistic society. Not only are different views held on *particular* ethical issues, as may well be the case in a society which is not pluralistic, but there are different understandings of life and the *bases* of moral judgments. Yet there must be some common morality that holds it together; and if informed judgments on ethical issues are to be made all aspects of them need to be explored. Ethical decisions are always difficult to make, but sweeping the issues under the carpet simply allows unexamined ethical attitudes to take priority, it does not find the right answer, or even the best compromise. This is a book of problems, and how various people have tried to overcome those problems. Like a book of worked examples in chess, it will not tell you how to win every game, but it might make you a better player.

Each of the small group of people who met to consider these issues brought his own expert knowledge and experience to bear on the problem. Some were scientists, whose daily work raised ethical conflicts or problems of conscience for which there are no simple solutions. Others were theologians or humanists: people with considerable experience and insight into the ways in which ethical problems can be tackled. And the problems considered ranged over the whole spectrum of science, both pure and applied: from the choice of research project to the publication of results, from the use of animals in experiments to the building of reservoirs, from the safety of foods to military research.

But, to begin at the beginning, how do ethical problems arise in science? When must a scientist's conscience temper those cherished commandments which, he believes, underpin the practice of science? To begin to answer this, consider one of the basic dogmas of science: the right to publish.

2 Science in the Open

> There is no truth existing which I fear, or would wish
> unknown to the whole world.
>
> *Thomas Jefferson*

> We have in our country three unspeakably precious
> things: freedom of speech, freedom of conscience, and
> the prudence never to practise either.
>
> *Mark Twain*

Freedom of inquiry, to seek and discover the truth of the world about us, and freedom to publish the results of one's research, are regarded by many scientists as absolutely fundamental to the business of science. And yet, in recent years, there have been several examples in which scientists themselves, as well as the general public, have argued that, in certain cases, these freedoms should be restricted. Moratoria have been proposed in certain areas of research, such as genetic engineering, whilst in others a voluntary ban on publication has been suggested. In turn, these developments have stimulated vigorous defences of these basic rights of the scientist. Professor Eysenck who, as he puts it, 'takes his stand with Thomas Jefferson', has recently published such an article, staunchly defending the right of the scientist to publish his results (Eysenck, 1975). Who is right?

Professor Eysenck's paper opens as follows:

> It used to be taken for granted that it was not only ethically *right* for scientists to make public their discoveries; it was regarded as their *duty* to do so. Secrecy, the withholding of information, and the refusal to communicate knowledge were rightly regarded as cardinal sins against the scientific ethos. This is true no more. In recent years it has been argued, more and more vociferously, that scientists should have regard for the social consequences of their discoveries and of their pronouncements; if these consequences are undesirable, the research in the area involved should be terminated, and the results already achieved should not be publicized (Eysenck, 1975, p.23).

Professor Eysenck does not agree with these recent views, and argues that they are not only mistaken, but may, in certain cases, be positively dangerous. Much of his discussion is concerned with one particular case, and this can be considered in a moment. First, however, it will be useful to consider the general question. Has the scientist always both the right and the duty to publish his results? Is Mark Twain's epigram merely clever wordplay, or does it contain the elements of wisdom?

Suppose, for sake of argument, that a scientist discovered a way to blow up the world using only string, cough mixture and ink (which, fortunately, is fairly unlikely), and decided to publish his results. He would be widely condemned, and rightly so; the odds are that some psychopath would discover the method and, in a subsequent fit of pique, blow us all to pieces. This, of course, is an extreme and slightly absurd example. But only slightly absurd. Some modern methods of biological warfare, not to mention the 'bomb' itself, are equally lethal. In a world where organized terrorism is rampant, the case for secrecy in such matters sounds very convincing.

It is useless to suggest that the problems can be avoided by stopping such research; we already have such knowledge, and more may be discovered by accident. The issue is not one of preventing research, but of preventing the widespread communication of 'dangerous' knowledge. It is equally useless to argue, as Eysenck seems to do, that since the scientist cannot know with any certainty what the consequences of publishing his results will be, then publication is justified. To be sure, he cannot know in advance that someone will use his results to blow us up, but to ignore the possibility, to 'publish and be damned', is mere irresponsibility.

However, even if secrecy can be justified in certain cases, it should be avoided if possible. We live in an open society, one in which secrecy is tolerated only when it can be justified for higher social goals: protecting the rights of the individual, or protecting the nation from a potential aggressor. Even then, the pressures against secrecy are strong, and a vigilant press is quick to point to any mis-use of secrecy which they suspect. Moreover, quite apart from any idealistic dislike of secrecy, there are good practical reasons for the free and widespread communication of scientific knowledge, as Eysenck rightly points out. Social policy must be based on fundamental ethical ideals, but it must also be based on a sound knowledge of facts as revealed by science; social policy based on prejudice and ignorance is unlikely to succeed.

In most cases where the state deems that secrecy is desirable, the scientist will have entered into a contractual agreement not to publish his results. This applies, for example, to much of the research sponsored by defence departments. The ethical problems which can arise from enforced secrecy of this type form a separate issue, and will be considered elsewhere. The problem under discussion here is this: in what circumstances, if any, should a scientist withhold his results from publication, assuming that this freedom is not fettered by contractual obligations? And are such decisions a matter for the conscience of the individual, or should society and its institutions recommend, or even enforce in some cases, suitable guidelines? What, in short, constitutes a 'good case' for secrecy; what criteria should be used?

The issues involved can be drawn out by considering one notorious recent case. In 1969, A. R. Jensen, Professor of Educational Psychology at Berkeley, published a paper entitled 'How Much Can We Boost IQ and Scholastic Achievement?'. This has been reprinted in a collection of his papers, together with an account of the stormy reception it received (Jensen, 1972). Two years earlier, the Colman report, which had been commissioned by the US Government to evaluate the success of its compensatory education programme, an attempt to boost the IQ of children with significantly less-than-average IQ, had published its findings. These were, in short, that compensatory education seemed to have failed. Jensen's paper, a long and scholarly review on the hereditability of IQ, suggested reasons for this, and proposed a possible theoretical basis on which future programmes might be constructed. He included racial differences in IQ only as a part of his overall thesis, and, anyway, the editorial board of the journal in which his article was published had specifically asked him to include 'a clear statement of your position on social class and racial differences in intelligence' (ibid., p.11). But it was this part of his paper, less than 5% of the whole, which provoked the storm.

Previous research had clearly indicated that the average American Negro has a significantly lower IQ than the average member of the white population (Shuey, 1966). Jensen, in this article, attempted to find reasons why this should be so. He gave a careful review of the evidence as to the causes of this difference, whether it is caused by genetic factors or by environmental ones, and drew the following conclusions.

> All we are left with are various lines of evidence, no one of which is definitive alone, but which, viewed all together, make it a not unreasonable hypothesis that genetic factors are strongly implicated in the average Negro-white intelligence difference. The preponderance of the evidence is, in my opinion, less consistent

with a strictly environmental hypothesis than with a genetic hypothesis which, of course, does not exclude the influence of environment or its interaction with genetic factors (ibid., p.163).

This is not the place to enter into the battle, which has raged ever since, over whether Jensen's hypothesis is right or wrong. Nor is it necessary to discuss in detail the many technical questions which are involved in this debate, such as whether IQ and intelligence are the same thing or not. However important such discussions may be, they are not relevant to the problem under discussion here, which is whether or not Jensen was ethically right to publish his hypothesis in the first place.

The free and unfettered publication of scientific work is important for two reasons. First of all, it is an essential stage in the scientific method. One person's observations and hypotheses can become established scientific knowledge only by airing them for open discussion by the whole scientific community; without such discussion, science cannot advance. Secondly, public policy must be based on the fullest scientific knowledge that is available; policies which, through ignorance, run counter to the facts can never be permanently successful. Free publication of scientific knowledge, most scientists at least would argue, is an essential prerequisite to sound policy-making in a civilized society.

Jensen's paper served both functions. In seeking to establish the cause of the Negro-white IQ difference, he put forward a hypothesis, based on all the available evidence, for debate by his fellow scientists. Recognizing that the issues which he raised were of considerable social concern, he suggested ways in which educational policy could be improved, should his hypothesis prove to be correct. The subsequent issues of the same journal contained six articles by leading experts in reply to this paper, together with a further commentary by Jensen, so that, whilst Jensen himself has criticized the editorial policy as being biased against him, and only publishing articles which disagreed with his views (see Jensen, 1972, p.27), nevertheless some attempt was made to present these hypotheses and suggestions in a context of explanation, giving the whole spectrum of scientific opinion.

Despite many attempts, no one has yet succeeded in discrediting the arguments used by Jensen so that, by this test alone, it emerges after more than seven years as a piece of genuine scientific scholarship. The key observations which Jensen reports, namely the Negro-white difference in IQ, and the high hereditability of IQ within groups, were already widely reported in the literature, and both are accepted by wide sections of the scientific community (see, for example, Thoday, 1973). It is true that this acceptance is not universal, and that the debate on these observations still continues. In October, 1976, a number of criticisms, of a somewhat sensational nature, were made in *The Sunday Times* on some of the original research into the hereditability of IQ, to be followed, almost inevitably, by heated discussions both in the scientific literature (Schwartz, 1976; Wade, 1976) and in the correspondence columns of *The Times*. But the fact that observations are not universally accepted and are still under dispute does not, of itself, make it unethical to base hypotheses on those observations. And the inference which Jensen draws is clearly labelled as a hypothesis; Jefferson's dictum is irrelevant in this context. However, not only has this hypothesis been hotly disputed by other scientists, but Jensen has been attacked for even suggesting it in the first place, so that, in this sense, Jefferson's dictum hits home. Here is a possible fact which such scientists hope is false; but they would prefer not to have to test the possibility that it may be true. Such a denial of the scientific method might be justified if the social risks of airing this hypothesis for debate, and accepting the possibility that it may be true, far outweigh the social benefits which might arise from its discussion, though, even then, some scientists would argue that the search for truth is an independent duty.

What are these social benefits? Jensen, and others, have argued that ignorance of the causes of the Negro-white IQ difference is a major hindrance to sensible educational policy. However, many would disagree. 'Innate differences in ability and other individual variations should be taken into account by our educational system. These differences must, however, be judged on the basis of the individual and not on the basis of race' (Bodmer and Cavalli-Sforza, 1970, p.29). As Bereiter has pointed out (Bereiter, 1970, p.298), education deals with individual children of unknown genetic potential; moreover, the teacher can modify only the environment, so he must do the best he can, regardless of any genetic factors in IQ. However, this does not necessarily mean that these hypotheses are irrelevant to educational policy as a whole. If, as has been suggested, any environmental enrichment policy of the type currently practised in the USA is unlikely to increase the average IQ of Negro children relative to white children, which, like many other statements in this field, is widely disputed, then any policy of streaming in schools which is related to IQ will inevitably partially segregate children into black and white, whilst any policy of enforced desegregation will mix children of differing abilities and scholastic potential. If true, this fact has profound implications for educational policy. To quote Bereiter again 'The possibility that cultural differences are related to heredity, however, adds force to the need for schools to come to grips with the problem of providing for cultural pluralism without separation or segregation. This may well be the major policy problem facing public education in our time.'

Since Jensen's paper had a clearly defined purpose, both in increasing scientific knowledge and improving educational policy, why has it been suggested that it should not be published? The answer, the case against publication and indeed against any further research in this field, is very simple. 'In the present racial climate of the US, studies on racial differences in IQ, however well-intentioned, could easily be misinterpreted as a form of racism and lead to an unnecessary accentuation of racial tensions' (Bodmer and Cavalli-Sforza, 1970, p.29).

In all his writings, Jensen has stressed that he finds racialism abhorrent; individuals must be treated, in his view, as individuals, and not as members of a certain race, nationality or social class. There is no reason to doubt his sincerity in such statements. Unfortunately, problems arise, not because of what Jensen thinks or says, but because of how other people react to what he says. He has been accused of racialism, and his theories described as racist; indeed, one article, written in response to his own, was entitled simply 'How Racists Use "Science" to Degrade Black People' (Rowan 1970).

Part of this problem arises from the special nature of scientific knowledge. In this chapter so far, and indeed throughout the rest of this report, the phrase 'scientific knowledge' has been and will be used as if it were easily defined and readily understood. This, however, is not the case. The concept of scientific knowledge covers a wide spectrum, from established facts to reasonably certain hypotheses, and the degree of certainty associated with that knowledge varies along that spectrum. Indeed, 'rightly understood, science can point out to us only probabilities of varying degrees of certainty' (Glass, 1966, p.76). This does not invalidate the common sense view of knowledge: a statement like 'the earth goes round the sun' is sufficiently certain that any doubts can safely be consigned to the realms of philosophical speculation. But the scientific conception of knowledge is richer than this simple, common sense view, and it is one which the non-scientist may not always fully appreciate. One cannot assume that the public will know how to evaluate scientific knowledge, nor how to assess its social consequences, and failure to spell these out on the part of scientists may result in public misconceptions and become, for that reason, morally reprehensible. For example, the high hereditability of IQ within groups would mean, as one might suspect, that IQ is determined more by genetic factors than by the environment. But this would not mean that children will have the same IQ as their parents or siblings, nor would it necessarily mean that IQ cannot be altered by environmental or biological manipulation, for instance by the use of drugs (Eysenck 1973). And it would not necessarily imply that the difference in IQ between the Negro and white populations in the US are due to genetic factors.

These facts were spelt out in Jensen's original article (ibid., pp.114–121) and in later accounts (Eysenck 1971, 1973). Despite this, misconceptions arose in people's minds, and social harm resulted. Indeed, this is the root of the Jensen dilemma. The evidence suggests that Jensen was fully aware of the sensitivity of this topic, and took all reasonable care in writing his article and offering it for publication. In such circumstances he cannot be blamed for the consequences of his publication. He cannot be held responsible for the publicity given to his article by others, nor for the ways in which a minority of journalists chose to distort his hypothesis. And yet, if the article had never appeared, then these unfortunate results would have been avoided (though at a cost).

Jensen was faced with two possible courses of action. On the one hand, he could withhold the article, thus avoiding any social unpleasantness but, at the same time, thwarting the scientific method. Alternatively, he could take the risk and publish, knowing that benefits might accrue from publication, but also knowing that there were risks. The publicity arising from the article might lead some people to believe, wrongly, that all Negroes are mentally inferior to whites; the resultant controversy might, as Eysenck suggested in his 1975 paper, have the effect of reducing standards and increasing rancour within the social sciences. Such dangers might have been foreseen, as even a brief glance at the history of this subject will show (Provine, 1973). Recent events have reinforced the point. In his book, *Sociobiology: the new synthesis*, Professor Wilson of Harvard describes the new discipline which is 'attempting to explore the genetic contribution to patterns of human behaviour' (Wilson, 1976, p.342). As such, it has been attacked by extremist groups as yet another example of 'developing theories in the name of science which could legitimise the massive social inequity generated by Western capitalism', and dismissed on the grounds that 'genetics has as little to tell us about human societies as nuclear physics has to tell us about genetics' (BSSRS, 1976). The echoes from the Jensen debate are clear enough.

But the question remains. Should Jensen have published his article? Some people have praised his courage in publishing his paper, thus reasserting the basic duty of scientists to communicate their knowledge even if that knowledge may be socially unpalatable. Others have argued that he should not have done so, in view of what has happened since. Jensen himself, and others quoted by him, have argued that if the publication of his article had been handled somewhat differently, then the storm would not have blown up in the first place, and he would have performed a considerable service in postulating reasons for the failure of recent educational policies in the USA, and suggesting ways to improve the situation (ibid., pp.23ff.). The decision to publish was taken by Jensen and by the editorial board; whether one agrees or disagrees with this decision is a matter of personal judgment. Either way, some good, and some harm, would result from one's decision, as from almost any other decision.

What conclusions, if any, can be drawn from the Jensen debate? The most obvious lesson is that the issues involved are not always simple. Withholding results from publication inevitably makes it more difficult to arrive at the truth and thwarts the scientific method. Equally, it smacks of paternalism; a privileged minority, be it a single scientist or group of scientists, are taking on themselves the right to judge what it is in the best interests of their fellow men to know. Whilst this might be necessary in a few cases, in general restriction of the freedom to know and to debate are not in the best interests of a free and democratic society.

On the other hand, publication of scientific results and hypotheses, even in the most obscure scientific journal, is a social act, which may have undesirable social consequences. The scientist cannot ignore these consequences. People attach greater weight to the pronouncements of scientists than to statements by the average citizen, and the scientist has a duty to take reasonable care that his statements are not so liable to misunderstanding that social harm results from them. Ravetz has put the point succinctly. 'To claim the right to utter unsettling doctrines, while denying any responsibility for their effects, comes perilously close to the arrogation of power without responsibility, a traditional variety of immorality' (Ravetz, 1971, p.390).

Whilst it is neither possible, nor necessarily desirable, to draw up rules and regulations to deal with such problems, it is possible to outline principles which should be followed.

First, the scientist has both the right and the duty to publish the results of his research. Most members of the study group believe that there must be exceptions to this rule, but that these exceptions must be kept to a minimum.

Secondly, it is almost part of the definition of scientific knowledge that one person's beliefs cannot be taken for truth unless they are shared by others. For this reason, secrecy should not be the decision of one individual in isolation. The romantic picture of the dedicated scientist who, on discovering a dreadful fact, keeps it to himself and dies with his lips sealed is both unrealistic and dangerous. It is in the nature of things that, eventually, someone else will make the same discovery, and the dangers that the later discoverer may use such scientific knowledge to do harm are too great to risk. Society has evolved institutions which, under the protection of such provisions as the Official Secrets Act, can both shield society from the dangers of this knowledge and find ways to overcome these dangers, and it is to such institutions that the scientist must turn for guidance.

On the other hand, secrecy can be justified whenever publication would create a real risk to members of the public. No responsible person would advocate the publication of the string and cough mixture recipe given above. But one must always weigh the social benefits against the social risks. The banning of flick-knives is justifiable; the banning of carving knives is not. Similarly, many of the drugs used to alleviate suffering and prevent death are also highly poisonous; in such cases one does not ban the drug but restricts its use. Risks are involved, but the benefits justify these risks.

What of other cases, where knowledge or supposition is not dangerous in itself but may, through misunderstanding, distortion or over-reaction, lead to undesirable social consequences? Two points can be made. First of all, it is wholly wrong for society to make any attempt to put pressure on a scientist to revise or suppress any genuine scientific results, simply because they are contrary to particular political beliefs or social policies, no matter how sensitive these may be. Political censorship or coercion of that type is wholly undesirable. To argue that Jensen should have been prevented from publishing his paper, or even to suggest that pressure should have been placed on him to withhold it, because it seemed to oppose certain political beliefs, is totally unjustifiable. The right, and indeed the duty, to decide whether or not to offer the paper for publication must rest with Jensen alone; one may agree or disagree with his decision, but one cannot deny him that right, nor absolve him from that responsibility.

But even if the suppression of such results is wrong, their publication must be handled with considerable care. Everyone concerned in the process must bear some part of the responsibility: the author of the article, the institution which gives him his scientific credibility, the editors of the journal, the referees, and the journalists and broadcasters who disseminate these facts and hypotheses to a wider audience. For example, no author should attempt to publish socially explosive results in *The Daily Bugle* before they had been published in the scientific literature (though he may be in some difficulty if he has unwisely entered into a contractual agreement to give the newspaper the right of first refusal over publication). Similarly editors and referees should assess not only the scientific credibility of a proposed paper, but also its social consequences, and be prepared to discuss with the author a strategy for publication in contentious areas. In areas of doubt, editors should be prepared to call in more referees if necessary, to obtain a consensus of opinion, and, in the event of publication, commission other articles, to be published alongside the original, giving the views of other experts in the field. Every effort should be made to publish these results in a context of explanation, indicating their degree of probability, their bearing on policy, their relative importance, and dealing in anticipation with likely misunderstandings and misapplications. The active co-operation of the media should be sought in this process; the latter, for their part, should be prepared to eschew sensationalism and immediacy in favour of social caution and careful reporting of the truth. It is not always easy to encapsulate scientific conjectures in a few, pithy sentences; the journalist's task is not an easy one. However, if it proves impossible (and the journalist should not rely on his opinion alone but should consult both the original author and other experts in the field), then he should remain silent. Half a loaf in this case is worse than none.

The scientist's responsibilities do not end with the act of publication. If certain organs of the media misrepresent the facts, then he must use every means at his disposal to correct this publicly. He should be able to appeal to the scientific institutions which support him for help, which they should be prepared to give. This does not mean that such institutions should try to decide what is right or wrong; rather, it means that if an institution gives an individual scientific credence by bestowing upon him the title of professor or numbering him amongst its fellowship, it should be prepared to reinforce that credence publicly by using every means at its disposal to give him a public platform if the circumstances demand it; in short, to put its mouth where its money is.

None of these suggestions is particularly radical; indeed, many simply describe what is current practice. They represent the minimum guidelines necessary, and it is probably unwise to attempt to go further. One might envisage ways in which the scientific institutions might take a more active role in vetting possible papers in advance but, on balance, decisions of this kind are probably best left to discussion between responsible individuals; institutional control is a useful safeguard but only as a background resource.

If these guidelines had been followed, would the storm over Jensen's paper have been prevented? The answer is probably no. Indeed, the publication of this paper followed many of these guidelines quite closely (see Jensen, 1972). No system is foolproof, and disputes will always arise. But it is better to take the risk, with appropriate safeguards, than to hide behind secrecy and privilege.

The problems of publication have been explored in some depth because they provide an excellent example of how the values which scientists adhere to, neatly summarized in the Jeffersonian quotation, may come into conflict with other social values. But there is a wide spectrum of opinion on this topic, and many scientists would not accept some of the arguments used above, including one member of the Science and Ethics Committee. The position which such scientists adopt is the one with which this chapter opened: not only have scientists the right to make public their discoveries, but it is their duty to do so. The decision whether or not to publish a particular scientific paper, they would argue, must be based only on considerations of scientific merit. They would not advocate the publication of dangerous technical knowledge, such as the string and cough mixture recipe, but they would draw a clear distinction between this type of technical knowledge and pure scientific knowledge of the type published by Jensen. The latter, they would argue, must be published, regardless of social considerations; the fact that their fellow men may lack the maturity to live in peace with certain items of scientific knowledge, is, in their opinion, irrelevant. After all, if the topic of Jensen's paper had been athletic prowess rather than IQ, the fuss would never have occurred. They would accept that any strategy of publication should take social factors into account, but not that such factors should affect the decision to publish.

The real force behind this point of view is not just that a scientist has an absolute right to publish his results but that he has no right to withhold his results, whatever his motives for wishing to withhold them may be. It is a commandment: thou shalt publish the results of thy work. Later chapters will discuss this type of imperative in more detail; for the present it is sufficient to note that the majority of the Committee disagreed with this view, but that it neatly illustrates the deep divisions and differences of opinion which surround this subject. It is these divisions which may make this such a difficult subject (indeed, it caused more difficulty to the Science and Ethics Committee than any other topic covered by the report).

References

Bereiter, C., 'Genetics and Educability: Educational Implications of the Jensen Debate', in J. Hellmuth (ed.), *Disadvantaged Child, Vol. 3, Compensatory Education: A National Debate*, Brunner-Mazel, New York (1970).

Bodmer, W. F. and Cavalli-Sforza, L. L., 'Intelligence and Race', *Scientific American*, 223, no. 4 (1970), pp.19–29.

BSSRS, Science as Ideology Group of the British Society for Social Responsibility in Science, 'The New Synthesis is an Old Story', *New Scientist*, 70 (1976), pp.346–348).

Eysenck, H. J., *Race, Intelligence and Education*, Temple Smith (1971).

Eysenck, H. J., *The Inequality of Man*, Temple Smith (1973).

Eysenck, H. J., 'The Ethics of Science and the Duties of Scientists', *The Advancement of Science* (new issue) no. 1 (1975), pp.23–25.

Glass, B., *Science and Ethical Values*, Oxford University Press (1966).

Jensen, A. R., *Genetics and Education*, Methuen (1972).

Provine, W. B., 'Genetics and the Biology of Race Crossing', *Science*, 182 (1973), pp.790–796.

Ravetz, J. R., *Scientific Knowledge and its Social Problems*, Oxford University Press (1971).

Rowan, C. T., 'How Racists use "Science" to Degrade Black People', *Ebony*, 25, no. 7 (1970), pp.31–40.

Schwartz, J., 'After Burt, what's left?', *New Scientist*, 72 (1976), pp.330–331.

Shuey, A. M., *The Testing of Negro Intelligence*, Social Science Press (1966).

Thoday, J. M., 'Educability and Group Differences', *Nature*, 245 (1973), pp.418–420.

Wade, N., 'IQ and Heredity: Suspicion of Fraud Beclouds Classic Experiment', *Science*, 194 (1976), pp.916–919.

Wilson, E., 'Sociobiology: a new approach to understanding the basis of human nature', *New Scientist*, 70 (1976), pp.342–345.

3 Dumb Animals and Vocal Minorities

Whatever goes upon two legs is an enemy.
Whatever goes upon four legs, or has wings, is a friend.
George Orwell

A scientist cannot have total freedom to publish the results of his research. The number of cases where social considerations proscribe his freedom may not be large, but he cannot ignore this implied background of social responsibility which governs his professional conduct. Publication is not just the communication of knowledge; it is a social act with social consequences. But what of research itself? Can a scientist choose any area of research, can he use whatever techniques will produce the results he requires, or are these freedoms circumscribed in any way?

Obviously, there must be restrictions. Any research which would place other people in situations of risk must be severely restricted. Similarly, there may be technical or financial restrictions on a particular project which, as will emerge later, may imply some degree of social control.

Experiments which use live animals are a typical example where the pursuit of knowledge is restricted by ethical and social considerations. Here the scientist is constrained by law (the Cruelty to Animals Act of 1876), under which he must demonstrate that the anticipated benefit from a particular experiment outweighs the pain and suffering which it might cause to individual animals, and that these experiments will be carried out in accordance with the standards required. If the scientist cannot satisfy these conditions, then the experiment will not be allowed.

Clearly, the setting of standards, and the balancing of benefits and risks, involves an element of proportion and, inevitably, of compromise. The strong and highly vocal anti-vivisection lobby, which attracts considerable public sympathy to its cause, believes that the law as it stands is too weak, and that many experiments which receive official sanction at present should be condemned as cruel and causing unnecessary suffering to animals. The public debate which such groups serve to stimulate cannot be ignored by the scientist. For each experiment, a scientist will consider both the law and the dictates of his own conscience and decide accordingly. And yet, even if he obeys his conscience, even if he sticks to the letter and spirit of the law, he may still find his actions opposed by some members of the general public, and will be drawn, inevitably, into this debate. When is it justifiable to use animals in scientific research, and when is it unjustifiable? Should there be a total ban on such experiments, or should all be allowed, or is there some suitable compromise? If the last of these is the case, then is the law as it stands a true reflection of the consensus within society of where that compromise should be made?

Unnecessary cruelty to other living species which are capable of suffering has long been universally acknowledged as wicked. Of course, there are difficulties in defining suffering in this context. Chop a few leaves off a rubber plant and it will visibly 'suffer', in some sense of the word. Pain and suffering are essentially human concepts however, and, whilst their extension to other species is possible, it is not always clear exactly what they mean in such circumstances. One might object that this is merely a quibble; pull a cat's tail and it obviously suffers. But do fish suffer? Or bees? Or rubber plants? Or bacteria? Obviously there are differences, but where does one draw the line?

Fortunately, it is possible to bypass this difficulty. There is general agreement on those types of animal behaviour which should be interpreted as suffering, so that, even though one can never be sure of the extent to which that animal's experiences are comparable to human suffering, the evidence is such that we have a moral obligation to extend the concepts of pain and suffering to animals. But, even if most people can agree on when an animal suffers and when it does not, this does not explain why cruelty which causes suffering should be morally wrong. After all, animals suffer considerably in nature, as when African Hunting Dogs eat their prey alive, or lemmings commit mass suicides in times of overpopulation.

One answer to this problem is that cruelty to animals is regarded as a moral defect in human beings. Cock fighting was banned because people found the sadistic pleasure which the spectators took in the sport morally disgusting. Similar feelings may motivate those who wish to ban fox hunting, or even boxing. Whilst there is more than a hint of puritanism in this attitude (though, in fairness to the puritans, scrupulosity might be a better word), it is nevertheless one which is prevalent in this country. Another possible answer is that many human beings identify themselves with animals so that, when the animal suffers, the person suffers as well, and the only way to alleviate the human suffering is to remove the animal suffering which causes it. Some people may dismiss this as over-sentimentality which should not be indulged, but such an attitude does not make the degree of suffering any less.

Underlying these arguments, however, is a much more basic attitude. This is the principle of respect for sentient life: a desire to avoid cruelty to animals, whether it arises from the man-regarding considerations of the previous paragraph, when cruelty brutalizes man himself, or from religious belief, or simply from the high value placed on the well-being of animals. Like the principle of respect for human life, it is so basic to our ethical creed that it scarcely needs explanation. Where it differs is in its application. The incurably sick human being will be kept alive whatever his suffering; the animal, in similar circumstances, will be 'put to sleep'. This sharp contrast gives a clue to man's special attitude to animals.

Consider the accounts of ethologists – scientists who watch animal behaviour in the wild and have, as one of their basic rules, 'thou shalt not interfere' – which describe the difficulties they feel in watching animal suffering without attempting to alleviate it. They know that such suffering is both necessary and universal in nature, yet they cannot watch it without experiencing strong emotions. Their behaviour is irrational, but clearly illustrates how animal suffering can rouse a sympathetic response in human beings.

Should the ethologist be condemned for his non-intervention? He has two possible defences. Firstly, he is not actively involved, he is merely witnessing one example of natural laws. This is a valid but rather weak argument, which immediately runs into difficulties when one considers the role of the vet. We would condemn the latter if he sat back and let animals suffer, so why not the former? Fortunately for the ethologist, he has a second line of argument. It is this. If he intervened on every occasion when he witnessed animal suffering, then his ability to practise his craft would disintegrate. The scientist can only hope to discover how nature works by observing it from the outside. If he continually interferes with the processes of nature, then he can never be sure whether what he observes is a natural process or a result of his interference. Since the science of ethology brings many benefits, both to mankind and to the animals themselves, one can argue that the realization of these benefits is a higher social goal than the disruption of those natural processes which lead to animal suffering and that, for this reason, the ethologist is justified in his non-intervention. The vet, on the other hand, is in a totally different position. Whilst the ethologist's profession is to watch animals, the vet is professionally bound to intervene. His function within society is specifically to alleviate the suffering of individual animals.

At first sight, these two professional codes of practice seem to be incompatible. One seems to sanction the disregard of animal suffering, whilst the other seems to demand that every effort should be made to alleviate it, two apparently irreconcilable points of view. This dichotomy disappears, however, when one realizes that the prevention of animal suffering, albeit a valid ethical principle to which everyone should adhere, cannot take absolute precedence over all other values. Here, as elsewhere, there is a tension between different values and one must weigh the potential benefit to man arising from an animal's suffering against that suffering itself. In the case of a pet, its suffering brings little or no benefit to man, so that the vet is wholly justified in attempting to overcome it. The suffering which the ethologist witnesses, on the other hand, may be of considerable value to man, so that it is only in rare cases that he can justify his intervention (see Van Lawick, 1973 for a well-known example).

This simple example illustrates that the principle of respect for sentient life is tempered by two other broad principles upon which all of man's dealings with animals must be based. The first is the principle of utility. Animals are used by man, and bring benefit to him in a wide variety of ways. Some of these, such as food production, inevitably involve the killing of animals; others, such as certain types of scientific research or sport, place animals in situations of risk and involve some animals suffering. The counterbalancing principle, which prevents undue exploitation of animals, is that of minimum hurt: the duty of care not to inflict unnecessary suffering on an animal. Just as the uses differ, so do the ways in which the principle of minimum hurt proscribes those uses, and in each case the position of balance must be sought. There will, of course, be differences of opinion. To an aficionado, the suffering of the bull is justified, but to most other people it is not. Similarly, the vegetarian may argue that the desire to eat meat is insufficient justification for the slaughter of animals, though his carnivorous compatriots will disagree. Despite these differences, however, it is still possible to reach a consensus on most topics.

Some people may find this emphasis on the utility of animals slightly distasteful. The sympathetic emotions which many pets arouse in their owners can be very strong, so that to classify this as merely another use of animals can bring forth angry denials. But, however real, however strong these emotions may be, they remain totally different from the emotions between human beings. The privileged position of the pet arises because of the strong feelings of its owner, not because we respect the feelings of that particular animal more than any other. Of course, the two are intimately intertwined, but it is because the emphasis always lies on one side of the emotional equation, namely the human side, that one can speak of use in this context.

How do these two principles apply to scientific research using animals? For each experiment, the principle of utility raises two questions. What are the potential benefits to man which might emerge from that experiment? And are there any other, more acceptable, ways by which those benefits could be realized? The principle of minimum hurt, on the other hand, raises two rather different questions. What degree of suffering in the animal would that experiment cause? And what would its effect be on the experimenter; would it make him less sensitive, in the long run, to animal suffering in general?

Consider the benefits first. Many of the recent medical advances which have produced treatments or cures for human diseases have been based on research involving live animals. To ban such research would lead to increased human suffering and death, or at least to fail to alleviate it, which would be wholly unacceptable. The welfare of human beings must take precedence over the welfare of animals, even if it is necessary to cause suffering to animals in order to remove suffering from man. But this does not mean that anything goes. In all cases there must be a proportion between the benefit to human beings and suffering inflicted on animals. Painful experiments on animals can be justified, but only if the benefits from them are correspondingly large, and those benefits cannot be obtained in any other more acceptable way. The moral restrictions on research into rabies, for example, will be far less than those on research into the common cold; the former is a particularly unpleasant and, once the symptoms have developed, almost invariably lethal disease, the latter merely a nuisance. And even then, however great the anticipated benefits, there may be some experiments which cannot be justified under any circumstances.

It is also important to realize that benefits do not only accrue from research which is of obvious practical utility. Many opponents of vivisection try to distinguish between 'pure' and 'applied' research, and to argue that it is much harder to justify the use of animals for pure research than for research which has an obvious purpose, such as the treatment of a particular disease; a point of view which, as Sir Peter Medawar has pointed out, is unrealistic.

In terms of their ultimate relevance to mankind, the difference between research into protein synthesis and on malaria is a difference of immediacy, and in the degree of diffusion of their effects. The work on protein synthesis stands further from practical application than work on malaria, but its results illuminate almost the whole of biology and medicine; they illuminate normal and abnormal growth, regeneration, reproduction, the synthesis of hormones, the production of antibodies, the multiplication of viruses and bacteria – surely a big enough dividend for any scholarly investment. You may disapprove of all experiments using animals, but it is scientifically and medically ruinous to approve only those with obvious practical uses and to reprobate all others (Medawar, 1972, p.85).

The issue becomes more complicated whenever there are alternative means of producing the same benefits. Such alternatives fall, broadly speaking, into two categories: experimental, and social. Since they are totally different, these two groups must be considered separately.

The experimental alternative to the use of animals is the use of tissue cultures. In an important book, first published in 1959, W. M. S. Russell and R. L. Burch defined three 'principles of humane experimental technique': reduction, replacement and refinement. By replacement they meant the replacement of live animals, as far as possible, by in vitro or tissue culture methods. Obviously, such methods would be free from the ethical difficulties which surround the use of animals and have, for this reason, been the focus of interest for some of the more responsible groups, such as FRAME (Fund for the Replacement of Animals in Medical Experiments), which have campaigned against animal experimentation in recent years.

Quite apart from the ethical advantages, tissue culture methods are better than live animals for a number of simple practical reasons. They are more manageable, easier to quantify and control, often cheaper and in some cases more reliable for experiments than using whole, live animals. For these reasons alone they are slowly replacing in vivo methods in some areas of experimental medicine, such as immunology.

But there are difficulties. There are certain types of knowledge which, at present at least, can only be obtained by experiments on animals. It may be possible to reduce our dependence on animals, but for the present, and the foreseeable future, the use of whole, live animals in certain experiments, such as exploring the functions of a given gland, is essential (see, for example, Paget, 1975). Many opponents of animal experiments (for example, Ryder, 1975) either attempt to ignore this fact or disclaim its validity. This is not justified. In a perceptive review of Ryder's book, Dr Bernard Dixon, the editor of *New Scientist*, points out how the book's main message – the urgent need to find alternatives to animal experiments wherever and whenever possible – is obscured by its 'total inadequacy in dealing with the central issue of the continuing need for animals in crucial medical research' (Dixon, 1975).

The very specificity of tissue culture tests, which makes them so attractive from one point of view, limits their usefulness. If, for example, one wishes to examine the effect of a drug on, say, the muscles, then it may be possible to test it on isolated muscle cells in a tissue culture. But it is almost a truism that all drugs have all actions; almost none have that supreme selectivity of action which ensures that they affect only one part of the body. It is precisely because a cardio-vascular drug will be used on an individual who has a brain, and an anti-depressant drug will be used on an individual who has a cardio-vascular system, that all drugs must be vetted as safe for all actions which, at present at least, inevitably involves testing on whole, live animals. Similarly, if a drug was suspected of having certain behavioural effects – epileptogenic effects for example – then this could only be tested by injecting it into an animal to see whether or not it produced seizures. Animal experiments are not necessarily more reliable than tissue culture tests, though many would argue that these are more reliable at present, but they do complement tissue culture tests and provide an extra safeguard in the testing of new compounds. No coroner is likely to accept that every effort had been made to ensure the safety of a new drug if it had not been tested on live animals (unless, of course, such tests were banned by law). This emphasis on reliability also has another effect. At present, it has been argued that the only method available to assess the reliability of tissue culture tests is to correlate them with experiments on animals, so that increasing the former implies increasing the latter.

Another set of difficulties arises from the considerable experience which many scientists have in using animal experiments. In many cases, a scientist will know which tests to perform on which animals in order to ensure the efficacy and safety of a particular drug, but will not have the same depth of knowledge of what types of tissue culture tests to perform. In addition, he may well have far more ready access to facilities for animal experiments than to tissue culture facilities. These, one might argue, are spurious reasons; if the system is wrong, then it should be changed. Unfortunately, such changes will inevitably cost money and resources, and must be judged against all the many other desirable social changes which we wish to bring about. FRAME, for example, have proposed 'that tissue culture banks be established in all major research centres and that these banks should not only provide standardised cell strains and lines, but should also provide technical advice and expertise . . . more research funds should be made available to develop "alternative" techniques . . . and we would hope that more support for the development and evaluation of in vitro systems will be forthcoming from both the Government (via the Research Councils) and private enterprise' (FRAME, 1976). All this is very desirable, but would inevitably mean that someone else would get less money. It may be morally right to pay this cost, but this can only be assessed through comparison with other programmes of reform.

The social alternatives to animal experiments, where they exist, are much more contentious. If new surgical techniques, or new drugs or vaccines, are the only conceivable way to alleviate that particular example of human suffering, then it is relatively easy to justify experiments using animals to test such methods. If the human suffering is avoidable by some other means, however, then it is not so easy.

Smoking is an obvious example. We do not catch smoking in the way that we catch measles, nor are we born with it. Each person chooses whether or not to begin to smoke despite the warnings of its potentially harmful effects. If he becomes ill as a result, of course, then society has a moral obligation to try to alleviate his suffering. But, at the same time, society attempts, through moral exhortation, fiscal sanction and demonstration of these physical ill-effects, to deter people from smoking, albeit with only limited success. Short of banning smoking, which one can argue is neither practical nor, because it restricts the freedom of the individual, desirable in this country at present, then there is only one other alternative: to introduce less-toxic substitutes for tobacco. In this case, one must ensure that any substitutes which are introduced do not have effects which are equally harmful as those of tobacco. The only way to do this is by performing experiments on animals, and in particular by exposing whole animals to smoke.

This problem is made even more complicated by the fact that tobacco is addictive. There are basically three injurious elements in tobacco smoke: tar, the CO and the nicotine. Any substitute which removed the first two would be relatively safe, compared to tobacco, and would still induce dependence. However, any substitute which removed nicotine as well would also be safer than tobacco, but would probably never be used and would therefore fail in its objective of replacing tobacco and hence ameliorating the consequences of tobacco smoking. Any successful substitute must be safe, but it must also be used, and hence must contain the dependency inducing agent. The difficulty here is that many people would regard any experiment which sought to prolong this dependence, as the testing of such a substitute would undoubtedly do, as morally reprehensible.

Should animals be used in the testing of tobacco-substitutes, with or without the dependency-inducing agent? Clearly, society should use every justifiable means to deter people from smoking. But, if people continue to smoke, then equally, society should explore every justifiable means to ameliorate the harmful consequences, including the development of less-toxic substitutes. Since human suffering will occur if such substitutes are not introduced, or if they are not tested on animals first, then such tests can be justified, though one might demand that such experiments are more strictly controlled than experiments in testing new vaccines, and that a particularly painful test might be allowable in the latter case but not in the former.

The use of animals in the testing of new commercial and consumer products – cosmetics, paints, food flavourings and the like – is even more difficult to justify. Smoking is a problem which is already present within society, so that the ultimate justification for tests on substitutes is that it will alleviate human suffering which already, potentially, exists. In addition, the element of addiction in smoking clouds the issue of free choice; many a smoker will tell you that he would give up if only he could. But no one will miss the new hair dye that hasn't been introduced yet; no one will yearn for the new petrol additive that exists only as an idea in the mind of some industrial chemist. Unless, of course, the adverts tell him that he wants it.

This is the heart of this problem. If a manufacturer introduces a new product, then the chain of reasoning is fairly straightforward. Society demands as high a degree of safety in such products as is feasible and, what is more, demands that the manufacturer should be liable for any harm done by them. Recent legislation, such as the 1974 Health and Safety at Work Act, places increasing pressure on manufacturers to ensure the safety of their products and, at present, this safety can only be assessed by experiments on animals. The reason for such experiments, and hence their justification, is the same in this case as any other: the prevention of human suffering.

But go back a step in the argument. If a new product is introduced, then there is a social need for it to be tested on animals. But is there a real social need for that product in the first place? If there is no such need, then it might be more ethical not to introduce it. People want better deodorants, tastier foods, and easy-to-apply paints, but it is hard to argue that there is a real social need for such goods, particularly since they need so much advertising to make them commercially successful. As an ethical priority, the satisfaction of such needs must rank pretty low. Then again, one might argue that the welfare of our society is founded on the success of industry and commerce, and to inhibit the development and marketing of new products would stifle the progress that they need to survive and prosper. But this is a matter of political judgment. One might just as easily argue, depending upon one's point of view, that the only real benefit is to line the shareholders' pockets still further.

Obviously, one cannot ban the introduction of all new products. Equally obviously, in many cases, the social needs which they fulfil are so unimportant that the tests which would be acceptable for a new drug would be wholly unacceptable for a new cosmetic. This does not necessarily mean that society should accept a greater degree of risk from such products. This might be one solution, but it would be unacceptable if there was any risk of another tragedy on the scale of the thalidomide affair. A better solution might be to demand as high a degree of safety in new products as is now required, but to inhibit their introduction by a manufacturer by making it more difficult for him to obtain permission to undertake the necessary tests: in short, to strengthen the law.

1976 was the one hundredth anniversary of the Cruelty to Animals Act which governs the use of animals in research in this country. Not unnaturally, considerable discussion of the efficacy and relevance of the Act to present circumstances took place as a result, much of it pitched in highly emotive language. A memorandum submitted to the Home Secretary by Lords Platt and Houghton, for example, baldly states that 'present control under the Act is inadequate. The catalogue of mutilation, misery, downright cruelty to animals in experiments is both lengthy and horrifying' (quoted in Vine, 1976). The Research Defence Society's response to this memorandum was described by one author as containing 'misrepresentation, non sequiturs and arrogance in about equal proportions' (Dixon 1976), whilst another author attacked the original memorandum as 'a report of unsubstantiated platitudes', displaying 'unscientific language', 'obvious bias, not least political', 'inexactitudes' and 'contemptible efforts at denigrating' (Vine, 1976). Evidence of strong feelings, perhaps, but not the sort of reasoned discussion which might improve the situation.

Many responsible scientists believe that the law as it stands provides adequate safeguards. Certainly, the situation in this country is far better than in the United States, where the absence of similar legislation inevitably means that many cruel experiments occur. Others, however, have voiced genuine worries. Since the Act is over 100 years old, and was designed to control a few hundred experiments a year and not the 5½ million which now take place, might it not be in need of review? It does not necessarily follow that it is out of date (any more than the vast increase in world population necessarily means that the Ten Commandments are in need of an overhaul) but it would still seem timely to have a fresh look at the situation. Other doubts have been expressed on the lack of control over examination of, or even training in, the necessary technical skills which a research worker must possess. Others have suggested that new legislation should 'distinguish between the medical and non-medical purposes of vivisection and (to) discourage the latter' (Whitehead, 1975).

No law can be perfect, particularly one which attempts to govern a complex ethical question like vivisection where there are deeply-felt convictions on both sides. But it is unfortunate, and perhaps inevitable, that rational discussion of such an emotive subject can so easily be submerged beneath rhetoric and propaganda. It is all too easy to produce photographs of animal experiments which will inflame public opinion, but it only serves to distort the argument. If the law does need to be reviewed, one hopes that the reviewers can insulate themselves and that their decisions will be based on calm and rational discussion, and not on some of the more frenzied outpourings which seem to dominate the vivisection debate.

References

Dixon, B., 'In the Name of Humanity', *New Scientist*, 65 (1975), p.465.
Dixon, B., 'Think Again, RDS', *New Scientist*, 71 (1976), p.370.
FRAME, 'Response to the Minutes of a meeting of the British Association Study Group on Science and Ethics', unpublished.
Lawick, H. van, *Solo*, Collins (1973).

Medawar, P. B., *The Hope of Progress*, Methuen (1972).

Paget, G. E., 'The Ethics of Vivisection', *Theology*, 78 (1975), pp.355–361.

Russell, W. M. S. and Burch, R. L., *The Principles of Humane Experimental Technique*, Methuen (1959).

Ryder, R. D., *Victims of Science*, Davis-Poynter (1975).

Vine, R. S., 'Cruelty to Animals', *New Scientist*, 71 (1976), p.588.

Whitehead, P., *Hansard, House of Commons Reports*, London, 9 April 1975.

4 To Heal the Sick . . .

Thou shalt not kill; but need not strive
Officiously to keep alive.
Arthur Hugh Clough

Any discussion on the use of animals in research inevitably strays into medical ethics. The ultimate justification for testing new drugs and other chemicals on animals is that, without such tests, the degree of human suffering would be considerably greater, and the relief of human suffering is accorded a very high ethical priority in almost all communities. But medicine itself is riddled with ethical problems which, precisely because they are directly concerned with the amelioration of individual human suffering, receive widespread debate and, at times, arouse passionate feelings.

There are a number of well-established methods for tackling such ethical problems as they arise. Individual doctors often take a great interest in such issues, no doubt because the personal character of much of their work, and the fact that they accept final responsibility for the well-being of their patients, puts greater moral pressures on them than on those whose business is primarily research. Although the formal teaching of ethics to medical students is still in its infancy, there are numerous professional and student groups in which ethical issues are discussed, and the professional associations themselves have occasionally laid down ethical guidelines for their members. (Examples from the Medical Research Council and The Royal College of Physicians are quoted in the references.) The recent formation of a Society for the Study of Medical Ethics, with its own journal, is evidence of this growing concern, and the establishment of ethical committees for the approval of clinical research projects is a further straw in the wind, even though these committees seem to vary greatly in their effectiveness. External agencies, notably the churches, have also traditionally taken an interest in medical ethics, and interdisciplinary discussion in this sphere is widely accepted as a valuable exercise. Several notable reports have resulted from such discussions.

It cannot be said, therefore, that there is any lack of opportunity for airing problems as they arise, but even so there is a place for individual initiative. One example of this is provided by an ad hoc interdisciplinary group which was set up in Newcastle to consider problems arising from the selective treatment of spina bifida and which published a report of its work in *The Lancet* in 1975, on which much of what follows is based. The topic was chosen because of its immediate importance to some of the medical members of the group, and because it pinpoints very clearly the kind of decision which has to be made in relation to several categories of deformed new-born babies. However, the group inevitably found itself drawn into a discussion of the general aims of medicine, which is typical of ethical discussion. Particular problems must be evaluated in the general context which surrounds them, and this widening of the discussion can be a valuable corrective to those who are immersed in their own specialisms.

Advances in surgical technique have made it possible to operate on most of those born with spina bifida, but the improvement to be expected in severe cases is not large and the quality of life enjoyed by the survivors is not high. In the worst cases, the chances of escaping severe disablement and a life in institutional care are negligible, but the alternative, without operation, is almost certain death. The question is, therefore, whether the doctors in charge of severe cases are right to withhold surgical treatment, knowing that the babies thus selected stand little chance of survival. This leads to broader questions. Have doctors the right to decide what kinds of life are worth living, or have they a duty to give their patients whatever treatment is available, even if it is only likely to cause a marginal improvement? The policy of selective treatment, as now widely practised, depends upon an appraisal of the results of early surgical treatment, first published in 1971, in which it was claimed that certain 'initial adverse criteria' enable a fairly accurate prognosis to be made soon after birth. Granted the existence of such criteria, are they a sufficient basis for a life or death decision?

Some doctors have no doubt that, in the face of such adverse criteria, it is right to relieve suffering even at the cost of destroying life. The suffering in question may be that of the child itself, its parents, other children in the family, or society at large, which has to bear the cost of institutional care. One writer put the point succinctly: 'My conclusion is that we should put first things first; prevention of suffering comes before preservation of life' (Slater, 1973). And such decisions are being taken. One recent study shows that the number of deaths within the first year of life, for children born with spina bifida, hydrocephalus or both, fell from 1105 in 1960 to 530 in 1970. Since then it has increased: in 1972 the figure was 654. Similarly, the number of operations on children with spina bifida fell from 200 in 1970 to 131 in 1972. On the basis of the evidence available, these authors concluded that 'this recent increase in these rates of mortality suggests that the doubts (that have been expressed both by doctors and by the public) about the wisdom of keeping infants alive however disabled have been carried into medical practice' (Weatherall and White, 1976, p.3).

Others are less certain that this choice between the two traditional aims of medicine is so simple. What might be the long-term consequence of creating a medical environment in which judgments concerning pain and happiness were the sole criteria for making decisions about life and death? Are doctors willing or able to take upon themselves the onus of deciding what condition or 'quality' of life is 'acceptable' for their patients? The present balance between the aims of medicine, though it creates conflicts and may be impossible to maintain under extreme circumstances, nevertheless acts as a safeguard against the subtle acquisition of powers over life and death, which hitherto the medical profession have neither wanted nor been granted.

Behind this discussion about the traditional aims of medicine, and, as it were, supporting them, lie two broad ethical theories, both of which have been found necessary in practice to complement and correct each other. A utilitarian calculation of the balance of happiness and pain generally provides the most rational means of assessing social priorities. In any workable form of utilitarianism the calculation cannot be made in each particular instance, but is embodied in general principles. Even so, the weakness of this approach is that it does not provide much safeguard for the rights of individuals, and it is on this level that a more absolutist type of morality, adhering to principles like that of the sanctity of life and respect for human personality, can provide an important corrective. Similar problems face the cost benefit analyst who tries to place a monetary value on human life. His calculations may provide valuable insights into those implicit valuations which society makes, and may be of considerable help in making investment decisions but, ultimately, he must recognize that, in certain circumstances, society is not prepared to assign a price to human life. This problem will be explored in more depth later.

A new-born child with severe spina bifida has little to put in the scales of utilitarian balance unless the sheer fact of its humanity is respected. No doubt in many cases such respect for its life will be outweighed by the potential misery the child might suffer and cause. But unless there is seen to be at stake a conflict of principles, not just a single principle, the gradual assumption of powers over life and death could become too easy. And the converse is also true. To assert respect for human personality, and so to preserve life at any cost without considering what in general makes human beings happy, may lead, and often has led, to unnecessary suffering for the sake of blind adherence to beliefs.

Stating the ethical dilemma in these very broad terms may serve to demonstrate how practical questions have roots which go deep into the very nature of ethics. In fact it is possible to think of an ethical problem in terms of a series of levels.

At the deepest level there are questions about ultimate aims and life-styles and values. In medicine some of these have been codified in the traditional aims of medical practice, and reasons have already been given to suggest that the balance should not be altered.

At the intermediate level the implications of these decisions about values have to be spelt out in general terms. What, for instance, is the 'life' which is to be so highly respected? Are there degrees and qualities of life so that it is possible to specify a minimum below which efforts ought not to be made to preserve it?

Very severe malformation of the brain might be used as a possible criterion on the grounds that in such cases there could be no development of personal relationships, and hence no apparent personal life. But any less stringent criterion would imply making a judgment about the kinds of happiness which are attainable by the deformed, and would be a hazardous business. Most people want to go on living, however miserable their existence. The healthy person might well say 'I would find life intolerable under such conditions', but someone whose only alternative to such a life is death would probably disagree. Are there any circumstances where society has the right to over-rule the wishes of the individual and withhold treatment? Most people would say without hesitation that it does not. And yet, in countries where there are insufficient medical resources to treat all patients with particular chronic ailments, such as chronic renal disease, society is already taking such decisions, albeit by default.

Another way of approaching the problem on this level is to ask whether there is an obligation on the medical profession to use all possible means of preserving life, whatever the circumstances. There are growing doubts about the wisdom of accepting such an obligation in the case of the old and those with debilitating disease, but the issues are not so clear when it is a newborn child's potentialities which are at stake. Roman Catholic theologians have developed the concept of 'ordinary/extraordinary means' in terms of which the reasonable stewardship of life does not oblige anyone to use extraordinary means for its preservation. This distinction is a relative one, which is evolving all the time. Nevertheless, given a case in which the use of modern techniques will provide survival for a child, but will only offer a future which is going to be substantially painful and/or unhappy, one can say that the techniques are extraordinary means. On the other hand, if they could offer a future which would be substantially happy *for him*, however little he might contribute to society from a utilitarian point of view, they would not be classed as extraordinary. But, here again, one must recognize that happiness is a relative concept and, unlike the case of the adult sufferer, one cannot ask the child itself for its view.

At its third, and most practical, level, ethical discussion must be concerned with detailed factual considerations as these apply to individual cases. Questions about the social consequences of increasing the number of children in institutional care, about the emotional consequences for parents of accepting or rejecting a deformed child, and about the effect on other children in the family, are all relevant at this level. It is also important to consider the way in which decisions about particular children are made, and by whom, and the consequences of involving the parents in this decision making.

Ethical discussion at any of the levels indicated does not make choices any easier. On the contrary, by exposing the many factors involved, it may complicate matters. But this can be an advantage. At least it ensures that the policy of selective treatment, which is probably at present the best and most humane approach to the problem, is kept under constant review. The danger in medicine, as in many other fields of activity, is to regard unavoidable temporary compromises as final solutions.

References

Medical Research Council, 'Responsibility in Investigations on Human Subjects' (1963).

Medical Research Council, 'Responsibility in the Use of Medical Information for Research' (1972).

Royal College of Physicians, 'Report of a Committeee in the Supervision of the Ethics of Clinical Research Investigations in Institutions' (1973).

Slater. E. 'Severely Malformed Children. Wanted – A New Basic

Approach', *British Medical Journal*, ii (1973), pp.285–286.

Weatherall, J. A. C. and White, G. C., 'A Study of Survival of Children with Spina Bifida', in *Child Health. Studies on Medical and Population Subjects no. 31*, Office of Population Censuses and Surveys (1976).

Working Party Report, 'Ethics of Selective Treatment of Spina Bifida', *The Lancet,* 1 (1975), pp.85–88.

5 . . . and Feed the Hungry

Allah is very merciful. He gives us a little rice to eat
almost every other day.

Muslim woman

A century of science has transformed medicine from near-quackery
to a highly efficient technology. The average Briton no longer needs
to fear measles or scarlet fever as killer diseases, and the scourge of
smallpox, if recent reports prove reliable, has been all but eradicated from the face of the earth. There will always be limits to our
abilities, however, and it is at these frontiers that ethical problems
arise. Some arise from technical ignorance. If it were possible to
cure spina bifida in all cases, then the ethical dilemmas discussed in
the previous chapter would largely disappear. Unfortunately, whilst
it is possible to keep many children alive who otherwise would have
died, it is not yet possible to remove the debilitating consequences
of the disease in all cases, and it is for this reason that doctors are
faced with an ethical choice. Other problems arise from expense.
Smallpox can be treated on a world scale because such treatment is
relatively cheap. However, there is little immediate prospect of
treating all human beings with chronic renal disease; the cost would
be prohibitive, so effective treatment remains the prerogative of the
relatively rich, either as individuals or as nations which can afford a
National Health Service.

Hunger and malnutrition are second only to physical suffering as
causes of human misery. Indeed, there is a close connection between the two; no population can enjoy full health if its diet is
inadequate. People are still dying from starvation in various parts of
the world, and it is natural that we should look to science to prevent
this, just as the application of science has greatly reduced human
misery caused by disease.

As in the case of medicine, the record of science as applied to food
production is impressive. The development of higher yielding
strains of plants and animals, of which the 'green revolution' is but
one example, the production of fertilizers to enrich the soil, and
chemicals to control unwanted plants and insects which would
otherwise decrease production, and the development of more
efficient agricultural methods and machinery, have all contributed

to an enormous increase in the world's food production. This has been complemented by an equally radical advance in the possibilities for the safe storage and distribution of food, and the establishment of standards for its chemical composition. At the same time, nutritional studies have enabled us, in principle at least, to define an 'adequate' diet for different categories of people, and to prevent specific deficiency diseases caused by lack of individual nutrients.

Of course, food is not merely a matter of nutrition. Man does not live by vitamin pills alone; he has quite a liking for chocolate as well. Some dietary preferences arise from deep-rooted ethical convictions; the almost universal condemnation of cannibalism, or the Jewish restrictions to Kosher food, for example. Others are more a matter of fashion, like the many products of milk – butter, cheese, yoghourt, buttermilk, kefir, koumiss and so on – whose use varies enormously from country to country. Here too, science has played a major role, in developing commodities which have attractive qualities of colour, flavour and texture, or in producing acceptable substitutes to more expensive foods: margarine, texturized vegetable protein or 'vinegar and its brown imitations' (Dixon, 1976, p.97). These developments need not be morally bad; every society and culture has prized certain articles of food for such attributes, and their enhancement can lead to an increase in natural well-being. But one must look at the problem in more detail, to ensure that this use of science is not deflecting valuable resources away from the amelioration of hunger and undernourishment.

One of the most pervasive human dietary predilections is the desire for animal food. Indeed, many anthropologists have argued that this desire was the single most important factor in the rapid evolution of man from his vegetarian forebears. Observations within individual societies, and between national communities, excluding such remaining hunter-gathers as there are left in the world, show a close inverse relationship between the proportion of total energy intake derived from vegetable foodstuffs and economic level (Bronson et al., 1973). In short, we all, or almost all, want to eat meat, but the rich can afford far more than the poor. This raises two moral problems, neither of which has a simple answer. Meat is not essential to an adequate diet, and may thus be regarded as a luxury even though the desire for meat may be rooted deep in human history. We have a moral injunction to provide our fellow man with the necessities of life, but what about the luxuries? Is it immoral for Western man to eat a surfeit of meat, denying it to his poorer cousins?

Quite apart from dietary likes and dislikes, meat is also an expensive food to produce. In order to produce a unit weight of animal food, the animal concerned must consume up to ten unit weights of feed, which is itself often edible by people. Because of this, the effective food intake of members of prosperous nations may be several-fold that of members of poor nations. Since some of the latter are starving, this, on the face of it, seems iniquitous. However, as Professor Mogens Boserup, Danish Representative to the United Nations Population Commission has argued, leaving aside moral considerations, reduced consumption by prosperous communities does not necessarily benefit those who are less prosperous (Boserup, 1975).

What view should the food scientist take on these problems? As has already been discussed, scientists have a duty to explain their discoveries to their fellow citizens, so that society as a whole can take social decisions in the light of the fullest knowledge available. Increasingly, however, scientists are either being asked for or, often on ethical grounds, are taking to themselves the responsibility to give opinions. To the doctor, of course, this is a familiar situation, but to many other scientists it is not, and the problems which it creates underlie much of this and the two subsequent chapters.

The food problem is an easy one to define. Many people are suffering from malnutrition, and some are starving to death. What should the scientist do about it? He may propose some solution to this problem, but his proposal remains an opinion, and the history of the past two decades shows that such opinions may be incorrect and lead to wrong decisions from which harm, not good, may accrue.

He might argue that we need more food. There has been general acceptance of the belief that, for the world as a whole, per capita food supplies have been dwindling to the point of a 'world food crisis'. The evidence, however, does not support this view. The remarkable increase in the total food production of both rich and poor countries over the last two decades shows that, for the world as a whole, the per capita supplies have not only been maintained during a time of rapid population increase but have actually improved. FAO forecasts up to 1985 imply that this trend is expected to continue. Of course, the facts underlying this simple statement are complex. We are all familiar with beef mountains and wine lakes, just as we suffer from apparent shortages of sugar and potatoes. The political and economic pressures surrounding food production and distribution are considerable – the fluctuations between pessimism and optimism, the measures taken in exporting countries to reduce production in order to maintain prices – these were summarized together with the available statistics by Poleman in 1975. The conclusions which seem to emerge are that local food shortages, and the hunger and malnutrition which goes with them, are not the results of a world food shortage, but reflect an imbalance in the distribution of food, which is a political, not a scientific, problem.

However, the assumption by scientists, and others, that one of the main causes of malnutrition and undernutrition is a world shortage of food has led them to encourage every expedient of science and engineering to maximize food production. The introduction of increasing numbers of large machines for cultivating, sowing and harvesting, for spreading fertilizer and controlling pests, together with the mechanization of animal husbandry, has been accompanied by a breakdown of the traditional social structure of rural communities. Large estates have replaced peasant smallholdings, increasing rural unemployment and accelerating the widespread migration of populations from the country to the towns. While scientists cannot be held to be solely responsible for this social phenomenon, nor have all its effects been to the disadvantage of the communities involved, harm has been done for which scientists, because of their overwhelming preoccupation with the technological means for increasing food production, must accept some responsibility.

This drive for more food has had other, potentially harmful, effects. In chapter 8 it will be shown that the 'green revolution', whilst it has provided spectacular increases in food production, has also lessened the potential of plant varieties to combat disease, increasing the risk of world-wide epidemics which might, in the long run, drastically reduce food supplies. Such possibilities are not wild speculation. At the end of the nineteenth century, the vines of Europe were spared decimation by disease only by grafting on vines of American origin which were resistant to the scourge. If such vines, which produced poorer wine, had not existed, the wine industry might not have survived. Other problems are raised by the worldwide clearing of forests to create fresh agricultural land. The ecological balance in such areas has been destroyed and, in some cases, disastrous floods have resulted, draining nutrients from the topsoil. But, more than this, some experts fear that a major reduction in global forest cover may lead to a decrease in atmospheric oxygen, which has profound implications for the survival of animal species, including man.

Other scientists, who are not directly concerned with food production, have argued that the solution is not more food but less people. Here again, scientific advocacy has had considerable effect on the growing opinion that there are too many people in the world, but such policies are opposed by other, deeply held, beliefs: religious convictions against the use of contraceptives; political policies to improve and stabilize society by increasing population in countries such as China; social emphasis on the size and responsibility of the family, and the need to care for the parents in old age. One might argue that such beliefs are irresponsible if a country cannot feed itself, but then what right has Britain, which imports a very considerable proportion of its own food, to say to other countries that they ought to be self-sufficient, and reduce their population if they are not?

Even if one accepts that there is enough food to feed everyone, at least at present, and the problem is merely one of redistribution, this does not remove the ethical difficulties. Prosperous nations, by including a high proportion of animal foodstuffs in their diet have, as has already been pointed out, an effective food intake which may be several-fold that of poorer nations. It would be possible to discourage the eating of meat, by taxation, for example. It might then be possible, though it would be far from easy, to use the agricultural resources this released to channel more food into the third world. But does not Western man have a right to spend his wealth as he chooses? Not an absolute right, of course, but a right nevertheless. And what of the farmers who earn their livelihood by raising sheep or cattle? Have they not the right to do so? Should we ban pets, thus releasing the food they eat for human consumption? Should we ban the use of fertilizers on lawns, golf courses and cemeteries, so that they can be used on agricultural land in the third world? Have we, in short, a duty to equalize the standard of living between the rich and poor countries? Would it be politically feasible to do so? The problems are endless; the solutions elusive. The food scientist who ignores these ramifications and presses for one, simplistic, solution, may, in the long run, do more harm than good.

Similar problems beset estimates of human nutritional requirements. The energy requirement, or how much an individual needs to eat, depends on the energy he expends in a day's work, as well as his physiological idiosyncrasy and the current opinion of what constitutes an ideal physique. 'Miss World' and 'Mr Universe' are very different ideals from the contemplative scholar or violinist. Even if one accepts that the individual's diet should be such as to allow him to develop to his full physiological potential, the assessment of what that is, the delicate dividing line between incipient obesity and 'subclinical' deficiency, is not easy to determine. Ideal consumption levels for protein, fat, minerals and vitamins are equally difficult to assess. Shifting scientific opinion on the level of protein required, opinion based as much on philosophical temperament as on experimental evidence and observation, constitutes a flagrant instance of the influence of scientists on society. In 1971, FAO drastically modified their estimates of the millions allegedly suffering from protein deficiency by accepting an estimate of individual needs one third lower than that which they had regarded as scientific truth in 1946 and 1952. In such circumstances, where estimates of national food supplies are statistically dubious, and knowledge of the necessary nutrient intake is almost equally incomplete, the scientist is faced with a difficult problem. Does his special knowledge, limited as it is, make it his duty to assume some responsibility for public policy? Indeed, does it require him to take on the role of an oracular authority, even though he can be mistaken and, thereby, bring about hardship and distress for which he must assume some responsibility? Or should he restrict himself to informing his peers of the facts as he sees them, and play no further part in guiding public policy? If the facts and predictions which he brings forward subsequently prove to be incorrect, to what extent does this involve him in guilt? 'Ignorance of the law', it is said, 'is no excuse' and therefore blameworthy. Under what circumstances, if any, is ignorance of the facts similarly blameworthy? As has already been pointed out, the tendency for society to blame someone if things go wrong is increasing, but this is not to say that this trend is justified.

Just as it is difficult for scientists to define an adequate diet, or to predict world food supplies, so also it is difficult to determine the amounts of toxic substances in foods which can be harmful. When certain cereal seeds are treated with organo-mercury fungicides, the food harvested from them is increased by up to thirty per cent. On the other hand, it also involves risks of contamination, both to people and to wild birds. Who should decide what level, if any, of such fungicides should be used, and by what criteria? Consider another example. Most people would agree that the taste of sweetness, though devoid of any nutritional significance, contributes to human happiness. Given this, how is the ethical balance to be struck between the very uncertain possibility of harm from consuming cyclamates, which are now banned, and the well-documented, but familiar and consequently ignored, damage from obesity, heart attack and dental caries from the ingestion of a corresponding amount of sugar eaten in substitution?

Obviously, science has a part to play in determining the degrees of risk. But this is not to say that the fixing of an acceptable level of such substances is a scientific decision. One committee of the AAAS, whose report will be discussed in more detail later, argued that when a scientist or group of scientists state that a particular risk is 'acceptable' or 'unacceptable', they are making a social judgment, and not a scientific conclusion; moreover, scientists must emphasize this distinction, or the public may attribute scientific authority to statements which do not deserve it (AAAS, 1965).

This statement has echoes elsewhere. The Institute of Food Science and Technology of the United Kingdom has attempted, in 1975, to formulate a code of ethical behaviour for food scientists confronted with such problems. Scientists are adjured to ascertain the facts and interpret them in an unbiased way without selecting those facts which support a predetermined conclusion, disguise a hazard or claim that one exists where it does not. 'Wholesome' food is interpreted as possessing five qualities: pleasing to those who eat it, of a composition complying with the law, uninfected, of proper nutritional value and incapable of doing harm. In elaborating the fifth criterion, the Institute accepts the fact that absolute safety is unattainable and that a cat may indeed be killed by stuffing it with cream, no matter how wholesome. But besides the possibility of harm arising from excess, the Institute recognizes the problems arising from hazards to special groups such as those with peculiar susceptibilities or allergies, and to other previously unsuspected dangers. To maintain such standards, food scientists are expected to adhere to 'good commercial practice' and to monitor their products to verify the effectiveness of their precautions. The Institute lays special stress on the ethical responsibility of food scientists to avoid providing ammunition to 'those who actively distort the debate'. Scientists are to fulfil this responsibility by avoiding the publication of inadequately supported conclusions, by not 'making apparently authoritative pronouncements outside the area of their particular scientific expertise'.

The fact that responsible bodies of scientists are trying to draw up codes of ethical conduct is an encouraging sign, if only because it demonstrates an awareness of ethical problems. But this does not remove all difficulties. Even if the scientist wholly restricts himself to objective, scientific statements, which may or may not be possible or socially desirable, these may still give rise to mistaken policies. It is not obvious that such scientists are any less blameworthy than their colleagues who actively encourage certain policies. And such policies often are mistaken. Over a period of a decade or more, efforts were expended to manufacture 'fish flour' intended to enrich the diet of impoverished communities in protein. The scientific, technical and financial investment in this project did not produce its intended benefits. Schemes for fortifying unsatisfactory diets with synthetic lysine have proved to be expensive and equally disappointing. Looking back over the history of the past three decades, Dr M. Ganzin, Director of the Food Policy and Nutrition Division of FAO, in the 1975 Sanderson-Wells Lecture at Queen Elizabeth College, London, was forced to conclude that the results of nutritional studies, and of nutritional measures based on their findings, have not generally been reflected in corresponding benefits to the communities involved.

J. Boyd Orr demonstrated in 1936 that the plane of nutrition in Great Britain is directly related to wealth. FAO have subsequently shown that this is true of other countries as well. Whilst the simple diet available to a poor nation – provided that it is not too poor – need not necessarily be nutritionally inadequate, there is no doubt that malnutrition is commonly an affliction of the poor. The indications are that the only way to overcome hunger and malnutrition in a community is to increase its wealth. The food scientist, trying to solve what he believes to be a scientific problem with scientific means, may well be wholly mistaken. Technological expertise is necessary but not always sufficient to the solution of social problems.

The problems of feeding the world are inordinately complex. Producing more and better food, which is the obvious contribution which science can make, is only one aspect. Population policy, the freedom of the rich to spend their wealth as they choose, the fallibility of scientists and the human misery which can result from their errors of judgment, the problems of massive medical aid which, in many cases, preserves the life of a newborn child and creates just one more undernourished mouth to feed – all these are equally important, and the food scientist ignores them at his peril. But what view should he take? What part should he play in shaping public policy, when science and politics become inextricably intertwined?

References

AAAS, 'The Integrity of Science'. A Report by the AAAS Committee on Science in the Promotion of Human Welfare, *American Scientist*, 53 (1965), pp.174–198.

Boserup, M., *Development III, No. 2*, (March 1975), United Nations, Rome.

Bronson, D. and Severin, B., 'Soviet Consumer Welfare: Brezhnev Era', *Soviet Economic Prospects for the Seventies*, US Congress, Joint Economic Committee, Washington DC (June 1973).

Dixon, B., *Invisible Allies*, Temple Smith (1976).

Institute of Food Science and Technology of the United Kingdom, Professional Guidelines, no. 1, *Wholesomeness of Food* (1975).

Orr, J. B., *Food, Health and Income*, Macmillan (1936).

Poleman, T. T., 'World Food: A Perspective', *Science*, 188 (1975), pp.510–518.

6 Man on the Moon

This Nation should commit itself to achieving the goal,
before this decade is out, of landing a man on the moon
and returning him safely to earth.
President J. F. Kennedy, 25 May, 1961

Space exploration is the greatest single scientific endeavour in the history of man. Not unnaturally, it has captured the public interest, and absorbed our resources, more than any other comparable branch of science. The excitement it has created is understandable; sending men to the moon must rank as one of mankind's most cherished ambitions, and its achievement as one of the great events of our time.

Not that it has been free of problems. Consider, for example, some of the issues raised by the sending of men into space. The profession of the astronaut is a hazardous one; already there have been several deaths and, even with the most stringent precautions, further accidents may occur. One might argue that astronauts are volunteers, and the problems raised are no different from those raised by mountaineering or motor racing. To what extent should society be paternalistic, and protect the individual from dangers to which he would willingly expose himself? Or, to put it rather more strongly, to what extent should society try to prevent an individual from demonstrating personal bravery in the service of society (a question which will recur in the discussion on transplant surgery)? Nothing we do is free of risks, whether they are financial risks, risks to life and limb or risks of contravening some ethical principle such as the wrongness of waste, and in any comparison of risks and benefits, the demonstration of courage must be included in the list of benefits. But, whilst it may tip the balance in certain cases, the opportunity for personal bravery cannot, of itself, justify the risks, so that one must still ask whether the advantages of using men, as opposed to only machines, outweigh the human risks involved.

What are these advantages? The main justification for manned rather than unmanned spaceflight is that man can do what machines cannot. He has a versatility, and an ability to respond to unpredictable situations, which no machine can show. These advantages were summed up by Dr S. Singer, who in 1969 was an official in the US

Department of the Interior, in the following factual, if slightly in-human way. 'Man is largely self-repairing and self-correcting, and can also repair equipment . . . he is already designed and manufac-tured, and his development seems more easy and natural . . . man is a low cost, low speed computer which can be mass produced by unskilled labour' (Singer, 1969, p.6).

Each manned spaceflight should be taken on its merits. If the objects of the mission are of value to mankind, and these objects cannot be achieved without using men, then their use may be justified. However, there are complications. In the same article, Dr Singer went on to point out that manned spaceflight fascinates the public in a way which unmanned satellites can never do (compare the publicity given to the recent Jovian flypast with that given to the Apollo–Soyuz linkup), and he argued that, for this reason, manned missions may be an essential part of any space programme. Space exploration is an expensive business, and no democracy can carry the financial burden without considerable public support. More-over, the public will more readily support something which captures their interest than something which does not, whatever its potential benefit to mankind as a whole. To quote Dr Singer again (writing in 1969) 'If we deliberately de-emphasize the manned programme, we may find that we have no space programme at all, manned or unmanned' (ibid., p.8).

An illustration of these difficulties is provided by the latest space 'spectacular', the docking in orbit of an American Apollo and Russian Soyuz spacecraft in July 1975. Officially, the purpose of this mission was to test a compatible mechanism for rendezvous and docking so that, in principle, a spacecraft from one country might be able to rescue the crew of a crippled spacecraft from the other. As such it has been widely criticized, and commentators have argued, quite cogently, that these objectives did not justify the expenditure of $242,500,000, by the USA alone on this project. These argu-ments have led more than one commentator to suggest that this mission owed more to unemployment in California, created in part by a reduction in the space programme, and President Nixon's determination to get re-elected in 1972 than to any real plans for space co-operation and rescue (Lewis, 1975; Valery, 1975).

However, even if these arguments are valid and the official reasons do not justify the expenditure involved, this does not neces-sarily mean that this mission should be condemned. It has had three other important consequences: it has helped to alleviate the unem-ployment problem in California and elsewhere; it has helped to maintain public support for the space programme as a whole; and it has helped to boost public morale in the USA at a time when other events have had the reverse effect. Moreover, in all probability, none of these three objectives could have been fulfilled if the mission had been advertised in these terms.

Similar examples occur in other parts of the space programme. Indeed, considering the huge sums of money involved, it is inevitable that the telling arguments for or against particular projects should be political rather than scientific. President Kennedy's announcement, quoted above, was not an endorsement of the considered view of the scientific community. Whilst the latter had considered manned landings on the moon, these had been regarded as a low priority, and not something to be undertaken in haste (AAAS, 1965, p.181). The President's decision was clearly based on political, not scientific, arguments. Dr Wiesner, the President's scientific adviser, made the following point when asked if he agreed with this decision:

> Yes. But many of my colleagues in the scientific community judge it purely on its scientific merit. I think if I were being asked whether this much money should be spent for purely scientific reasons, I would say emphatically 'no'. I think they fail to recognise the deep military implications, the very important political significance of what we are doing and the other important factors that influenced the President when he made his decision (quoted in AAAS, 1965, p.182; see also Mandelbaum, 1969).

As Friedlander has pointed out, if the space programme were to be stopped today, it is extremely unlikely that the annual budget devoted to it would be re-directed to other scientific projects (Friedlander, 1972, p.102).

But it is the early history of space exploration which really makes the point, a history which Sir Bernard Lovell has outlined with admirable clarity. From the end of the Second World War onwards, the United States had a clear lead over the USSR in rocket technology, simply because almost all the German experts and documents had gone west and not east. And yet, despite rapid advances in missile development, and despite advice and pressure from scientists and others, relatively little attempt was made to develop a programme to launch artificial satellites. Not only had the government given a low priority to this programme, but it placed major obstacles in its way: by insisting that it should not interact with or divert any effort from the ballistic weapon development, and that neither the main military rockets available, the Redstone and Jupiter missiles, nor the main launching facilities at Cape Canaveral, could be used to launch a satellite. Overnight, everything changed. It was not that the scientific arguments were suddenly more convincing, nor had the scientists suddenly become more persuasive. It was because, on 4 October 1957, the Soviet Union launched Sputnik I and launching artificial satellites became, overnight, a matter of national prestige. The impact was immediate. Three weeks later, in response to renewed requests from scientists, the Army were given 3½ million dollars and told to launch a satellite by 30 January 1958, less than four months after Sputnik I. They failed, but only by one day: Explorer I went into orbit on 31 January 1958. Within a year, the total United States budget for research and development had almost doubled, an increase of one per cent of the GNP; within four years, the total expenditure on space activities per year had risen to ten times its value in 1957 (Lovell, 1973, pp.23, 24). The policy of the mid-fifties had been completely transformed, but the reasons were political, not scientific.

Clearly, political thinking dominates the decision whether or not to undertake particular space projects. But this need not necessarily be a cause for concern amongst scientists. Any decision to allocate large sums of money to projects with a scientific content must remain a political one and, all too often, scientists who oppose the political nature of such decisions merely display their political naïveté. This does not mean that there are no problems involved. Can the scientist turn a blind eye to the seemingly spurious reasons which were used to justify the Apollo–Soyuz linkup? Should he merely accept the political decision of his overlords, or not?

Some of these problems were explored by a Committee of the American Association for the Advancement of Science, whose report was published in 1965 (AAAS, 1965). They expressed concern on two counts. Firstly, they were worried that the political motivations and pressures underlying the space programme were detrimental to the proper scientific evaluation of experiments in advance and that 'if the scientific community is subjected to pressure or blandishments designed to solicit research activity which conforms to the purpose of the space research program, the free pursuit of knowledge will suffer' (ibid., p.184). One can sympathize with this view, but if it is intended to imply that science can, or indeed should, be independent of political values, then it is clearly naïve. The pursuit of science cannot exist in a social vacuum.

Their second objection concerned various statements by which scientific advisory groups had justified the Apollo programme with non-scientific reasons, such as 'man's innate drive to explore unknown regions' or 'the pursuit of national prestige' (ibid., p.184). The Committee did not necessarily disagree with such statements, but argued that the professional scientific adviser should confine himself to scientifically objective statements, since he might lend an impression of scientific objectivity to any non-scientific statements which he might make. Here again, one has considerable sympathy with this view. The difficulties facing the scientist when he enters the political arena are considerable, as the next chapter will show. But, whilst the denial of an effective political voice to scientists may preserve the integrity of science, it is doubtful whether it is socially desirable. It is a common criticism of scientists that they ignore, or at best undervalue, social and political factors, and it is a failing which will not be overcome by stressing that the integrity of science should be maintained at all costs.

Because the space programme is both a scientific and a political activity, different sets of values can come into conflict and create problems. It is impossible to decide whether or not it is justifiable to send men into space without examining the space programme as a whole, and considering not only the potential scientific benefits which might accrue from it, but its political importance as well. These political problems are acute in the case of space research, since the major motivation for the exploration of space arises from military considerations. The rocket technology necessary for space exploration was developed for purely military reasons. Moreover, since the first satellite was launched, almost two thirds of the payloads placed in orbit by both the USA and the USSR have been for military purposes (Lovell, 1975, p.4). When Dr Singer argued that manned space missions are necessary to maintain public support for the space programme as a whole, it should be remembered that most of this space programme has a military, not a scientific, purpose, and that, without these military incentives, to develop the necessary missile technology for example, it is very unlikely that any space programme would ever have been developed. Nor do the scientific and military aspects of space exploration merely co-exist; they are closely intertwined so that it becomes impossible to separate the two. The basic technology is the same in both cases; the only difference is in the payload. Were it not for this fact, Explorer I could never have been launched within four months of Sputnik I.

Even the payloads themselves, and the results of space experiments, can serve both military and scientific interests. Satellites which measure the shape of the earth 'have revolutionised geodesy and many aspects of geophysics', investigations which 'will deepen our understanding of the physical properties of the Earth and may well have a significant practical effect in the possible correlations with earthquakes' (Lovell, 1973, pp.68, 69). These same observations have improved the accuracy with which an intercontinental ballistic missile can hit its target from an error of several miles to one of less than a mile. Similar remarks can be made about navigational satellites which, whilst they have allowed the captain of a ship in mid-ocean to calculate his position to within one tenth of a mile, much better than the old errors of several miles and a real help to safety at sea, were originally developed so that nuclear submarines could accurately deliver their missiles.

This symbiotic interdependence of the military and scientific uses of space exploration has a number of consequences. The discussion of cost-effectiveness considerations, for example, becomes very difficult. Only the rabid anti-technologist would assert that space exploration offers no benefits to mankind. Not only has it provided benefits which could not have been made available in any other way, from X-ray astronomy to accurate navigation, but, in many of the more direct applications, such as forest fire detection or accurate weather prediction, the economic savings which result may be greater than the cost of the relevant satellites (Lovell, 1973). However, the mere existence of benefits does not immediately tell us what proportion of our finite, global resources should be devoted to space research. Nor do statements like 'The Soviet Union spends as much on space as on housing' or 'Every year Americans spend six times more on alcohol than on space activities, and three times more on tobacco' (Lovell, 1973, p.85) really help very much. Spending too much on space may be wasteful and hence immoral, spending too little may be failing to make a necessary investment to improve the future lot of man, but finding the balance point is a political decision, based as much on an assessment of political priorities as on any real ability to predict what dividends will result from investment in space. There is no simple way in which one can compare the short-term benefits which the alleviation of world poverty and disease would produce with the long-term benefits which space exploration promises; both are important, and resources must be allocated to each, even if this allocation is relatively arbitrary. But even if different programmes could be quantified in this way, it is difficult to see how such figures should be applied to the space programme, since the basic technology of space is provided by

Defence Departments for military purposes. It is not possible to disentangle the military and scientific aspects of space exploration sufficiently to be able to cost them individually.

Whilst the military programme provides financial support for scientific space exploration, the latter supports the military programme in other ways. It lends a certain public acceptability to an overall programme which would otherwise be purely concerned with what many people regard as the slightly distasteful business of defence. This means that every scientist who participates in space exploration, whether he is directly concerned with military projects or not, contributes in some way to the overall support of the defence programme, which involves him in certain ethical choices which are absent from most other branches of research.

The pressure for results is also greatest in military experiments. The AAAS Report, referred to above, discusses two notorious examples of military experiments which were opposed by members of the scientific community, and their consequences. These findings are of sufficient importance to be outlined here.

Experiment 'Starfish' involved the detonation of a 1·4 megaton hydrogen bomb on 9 July 1962, 250 miles above the Pacific. The reason for the experiment was the military interest in the disruptive effects on radio communication of atomic particles produced by a high altitude nuclear explosion. The experiment was strongly opposed by some scientists, especially radio astronomers, who argued that it might cause large scale and long-lasting changes in a still poorly understood region of the atmosphere containing the Van Allen belts, which had only recently been discovered. In order to evaluate the likelihood of this possibility, the US Government called together a group of leading scientists, who concluded that the effects of this experiment would 'disappear within a few weeks to a few months'. Accordingly, the experiment went ahead. Several artificial satellites were extensively damaged by the new radiation belt, including Telstar, the first of the communication satellites, which was launched the day after 'Starfish' and was rendered completely useless by the radiation within a few months (Lovell, 1973, p.43). More important, perhaps, was the conclusion by McIlwain, in a review of the experiment, that it may be necessary to wait more than thirty years before its effects completely disappear. In fact, the conclusions of the Government's ad hoc committee, called together in haste and pressed by the military authorities to give a quick decision, had been totally wrong. Of course, there is no guarantee that, if this question had been exposed to the thorough, if slow-moving, processes of the scientific method, then the correct prediction would have been made. But it would have had more chance of success than the hastily undertaken and ill-judged review which did take place.

'West Ford' was a project to place a band of copper needles around the earth, to establish the feasibility of establishing a new means of global communication which could not be neutralized in time of war by high-altitude nuclear explosions. The experiment was strongly opposed by radio astronomers, who feared that the needles would disturb signals from outer space, but their opposition to the project was hampered by the degree of secrecy surrounding it. As in the previous example, the key issue was how long the effects of this experiment would last but, in this case, since information vital to this question was classified, the military authorities were able to shrug off criticisms by outsiders as ill-informed.

The first attempt on this experiment, in 1961, was nearly a disaster; the launching rocket went into the wrong orbit, and the hazard of a long-lived belt of needles, which might have interfered with future astronomical observations, was avoided only because of an instrument failure which prevented the dispersal of the needles. The second attempt, in 1963, had a number of 'fail-safe' procedures added, and was a partial success. Its effects have now largely disappeared. However, 'West Ford', like 'Starfish', shows the dangers of undue haste and military secrecy in experiments with uncertain, and possibly dangerous or undesirable, outcomes.

It is not only military experiments that can give rise to such problems. The political pressure in America to 'produce in space' can also lead to ill-judged ventures which, if this pressure were absent, would not be undertaken. For example, as early as 1963, some scientists had warned that giant rockets could contaminate the upper atmosphere sufficiently to upset the weather, the Ionosphere and even the life-protecting ozone layer. When Skylab I was launched, on 14 May 1973, the Saturn V rocket which carried it removed most of the electrons from the upper F layer of the atmosphere over a region 1000 kilometres wide. (These electrons provide one of the radio mirrors used for long distance radio communication.) This incident was not serious; the atmosphere began to recover after a few hours. It was, however, apparently unforeseen by NASA's engineers, despite the scientists' warnings (Mendillo et al., 1975).

Far greater hazards lie in the possible transportation of micro-organisms from one planet to another. If life of any form, or the potential to support life, exists on any of our near neighbours in the solar system, then the consequences of transporting living material from one environment to another could be disastrous. The history of the earth provides many examples where such transportation has wreaked ecological havoc: rabbits in Australia or malaria-carrying mosquitoes in Brazil, for example. Fortunately, the authorities involved are aware of these dangers, and elaborate sterilization procedures and quarantine regulations for returning astronauts have been developed. Even so, some scientists doubt the wisdom of sending terrestrial objects to land on Venus or Mars until we have a much clearer idea of the potential of these planets to support life. Their concern is not merely to ensure that no foreign organisms are returned to earth, but that the natural environment of these planets, and any life which it might support, should not be perverted by terrestrial organisms. Lederberg and Cowie first warned of the dangers of contamination which might result if rockets crash-landed on the moon or other planets in 1958, but this did not prevent the deliberate crashing of Lunik II on the lunar surface in 1959. Disturbed by this lack of caution, Sir Bernard Lovell was prompted to write, in 1962, that 'the impact of a terrestrial rocket on Mars or Venus, in the manner of Lunik II on the moon, would certainly be an unmitigated scientific and moral disaster' (ibid., p.80). But this warning too has gone unheeded, and spacecraft have now been landed on both, despite our ignorance of their capacity to support life.

These four examples highlight some of the problems facing the scientist involved in space research. Clearly, if he believes that a particular project involves certain dangers, then it is his duty to bring them to the attention of those with the authority to make decisions. Difficulties only arise if he disagrees with the decisions taken. He may wish to try to bring public pressure on the authorities to try to change the decision but he must remember that he is neither infallible, nor possessed of a monopoly of moral good, nor is he necessarily the best person to judge the relative weights of scientific considerations as against military or political ones. Scientific values are not absolute; any scientifically undesirable effects which an experiment may have need to be weighed against its possible military benefits. In the case of 'Starfish' and 'West Ford', too much emphasis was undoubtedly given to the military benefits, but this cannot be an absolute generalization. These difficulties are exacerbated by the fact that many scientists working in space research have signed contractual agreements or oaths of secrecy. In such cases, if he is unable to bring sufficient internal pressure to alter the decision, he must either accept it or break his oath. It is impossible to give a general ruling to cover such cases; each must be judged by the scientist according to his own conscience, but he must be prepared to accept the consequences if society as a whole disagrees with his decision. The scientist who is not directly involved is in an equally difficult position, since he cannot make objective judgements if certain key facts are withheld from him for reasons of secrecy. The arguments against unnecessary secrecy are strong, and military authorities are, on occasion, guilty of both undue haste and undue secrecy. But this is not to say that all secrecy is wrong.

The excitement of space research, and the considerable funding available for it, have, not unnaturally, attracted many scientists to work in this field. Because of this concentration of effort, there have been rapid advances in the relevant areas of science in the past fifteen years. But equally, as one might expect in any human endeavour on this scale, there have been problems, and science and scientists have not emerged totally unscathed. No area of science is so deeply involved in political and military pressures, with the possible exception of weapons research, nor illustrates so clearly that science is a social activity, and its aims are social aims which can conflict with other values. The extra dimension which space research brings to these problems, which is largely absent from food science, is that these social aims are not universally acknowledged as morally desirable. But the scientist who tries to avoid these problems by retreating behind a façade of scientific righteousness fails to fulfil his duties as a human being.

Why are scientists so shy of the real world? The obvious starting point in trying to find an answer to this question is to see what happens when a scientist deliberately emerges into the world of social values, and to study the pressures and problems which confront him there. Then, having discovered what he is trying to avoid, one can begin to suggest answers.

References

AAAS, 'The Integrity of Science', A Report by the AAAS Committee on Science in the Promotion of Human Welfare, *American Scientist*, 53 (1965), pp.174–198.

Friedlander, M. W., *The Conduct of Science*, Prentice-Hall (1972).

Lederberg, J. and Cowie, D. B., 'Moondust', *Science*, 127 (1958), pp.1473–1475.

Lewis, R., 'Space Age: End of Act One', *New Scientist*, 67 (1975), pp.265–266.

Lovell, A. C. B., *The Exploration of Outer Space*, Oxford University Press (1962).

Lovell, A. C. B., *The Origins and International Economics of Space Exploration*, Edinburgh University Press (1973).

Lovell, A. C. B., 'In the Centre of Immensities', *The Advancement of Science*, (new issue) no.1 (1975), pp.2–6.

Mandelbaum, L., 'Apollo: How the United States Decided to go to the Moon', *Science*, 163 (1969), pp.649–654.

Mendillo, M., Hawkins, G. S. and Klobuchar, J. A., 'A Large-Scale Hole in the Ionosphere Caused by the Launch of Skylab', *Science*, 187 (1975), pp.343–346.

Singer, S. F., 'Goals for Man in Space', *Proceedings of the Orbital International Laboratory and Space Sciences Conference, Cloudcroft, New Mexico* (1969), pp.4–14.

Valery, N., 'Detente in Orbit', *New Scientist*, 67 (1975), pp.136–139.

7 The Teesdale Affair

One should never put on one's best trousers to go out to
battle for freedom and truth.

Ibsen

Early in 1976, a Committee of the British Association published a
report on 'Science and the Media'. Amongst their principal conclusions, this Committee stated that 'there is a need for journalists to
distinguish the varying degrees of scientific certainty and for scientists to separate the assessment of scientific facts from their opinions
about the social repercussions of such facts' (p.4). Others have gone
much further than this. Thus the guidelines issued by the Institute of
Food Science and Technology, which were discussed in the chapter
on world food, state that scientists should avoid 'making apparently
authoritative pronouncements outside the area of their particular
scientific expertise', a view which the 1965 AAAS report on the
'Integrity of Science' clearly supports. This latter report makes its
point sufficiently strongly that it is worth quoting at some length.

> If a scientist, as an individual citizen, wishes to promulgate a
> particular political course, he is of course free to do so. However,
> in our view, when such advocacy is associated with his organised
> professional scientific activity, the political or social intent
> acquires a wholly unwarranted cloak of scientific objectivity. This
> tends to obscure the fact that the political issue, despite its associ-
> ation with science, is, like all matters of public policy, open to
> debate. Such action on the part of scientists is likely to inhibit the
> free public discussion of the issue, and delay the development of
> an independent judgement by citizens generally (p.185).

The danger that a scientist's opinions may masquerade as objective
facts, either because he himself confuses the two or because he does
not emphasize the distinction to the general public with sufficient
clarity, are real enough, as the last five chapters show. Not that
it is a problem within the practice of science, as the BA Committee
points out. 'The professional scientist' they emphasize 'acquires, by
experience, a characteristic attitude towards knowledge and opinion, as well as skill in rigorous analysis and precise delineation of the
circumstances in which statements are valid.' But it is a problem

whenever the social applications of science are discussed, and scientists are as guilty of misdemeanours in recent years as are journalists or politicians.

The trouble is that science and politics are very different games. If a politician had to restrict his speech to facts, he would rapidly end up speechless. His job is more to persuade people, and to produce social action, than to establish the truth by careful discussion. The way in which he handles fact and opinion and the relationship between them is, whether we like it or not, subtly different from that of the scientist. Once he has recognized this difference, the scientist faces two possible alternatives. He may say 'I will have none of your politics; I will tell you the facts, and you can do what you like with them.' Such a course of action may preserve his integrity, but denies him any possibility of influencing political events if he feels that mistaken decisions are made. Alternatively, he may enter into political discussions but, if he does so, he will rapidly come to realize that he may have to compromise what he regards as his scientific integrity. The ivory tower is no place for the scientist who wishes to bring about social change.

Conservation is a matter of considerable importance to many scientists. They feel that society, in striving for a higher economic standard of living, is giving far too little attention to conservation, so that valuable resources are being irreversibly destroyed. Naturally, they try to do something about this but inevitably find themselves drawn into the world of politics and political values.

The Teesdale Reservoir Controversy is typical of this type of problem. Between 1964 and 1967 botanists and conservationists fought a long political battle to try to prevent the construction of a reservoir on the headwaters of the River Tees at Cow Green in Upper Teesdale, County Durham. After very considerable publicity and protracted hearings involving select committees in both Houses of Parliament, the enabling Bill was finally passed and the reservoir was constructed.

As in any conservation issue, there were powerful arguments on both sides. Building the reservoir would flood a remote area of considerable natural beauty, irreversibly destroying a number of plant communities which were of special interest to botanists. On the other hand, building the reservoir elsewhere would take longer and be considerably more expensive, placing a much greater burden on those local industries which needed extra water and, inevitably, restricting industrial growth and the creation of more jobs in the area, whilst failure to build it altogether would have an even more profound effect on the economy of the region.

The arguments are familiar to any conservation debate, but the solutions are rarely obvious. The balancing of short-term against long-term gains, or the assessment of the relative value of different courses of action, is never easy, but, for the present, these issues will be side-stepped until the next chapter, which discusses the problem of conservation in general terms. In addition, detailed accounts of the Teesdale case can be found in the literature (see, for example, Gregory, 1971). What is of interest to the present discussion is not whether the reservoir should or should not have been built, but to show how scientists can find it very difficult to bridge the gap from science to politics.

Throughout the case, there was no disagreement between the promoters and the opponents of the Bill on the main facts. The reservoir scheme was designed to inundate about a square mile of upland country in a remote moorland area. On the eastern edge of the site, and partly within the (now) inundated area, there was a remarkable plant community containing an assemblage of nationally rare plants growing on a very unusual geological outcrop (the so-called 'sugar limestone'). Interspersed with this were gravelly flush areas with other nationally rare plants. A basic list of these rare species has been known to science since the early nineteenth century, though some less conspicuous members are much more recent discoveries. Some of this remarkable vegetation is now permanently submerged and destroyed.

Scientists, especially professional botanists, played an important part in the controversy, and indeed were partly responsible for the emergence of any organized opposition to the scheme when it first became known. Both the promoters and the opponents of the Bill called scientists as 'expert witnesses', to present the ostensibly objective facts, and the Nature Conservancy, which was the Government body directly concerned, gave such evidence. But even the Conservancy, however, was not neutral; after a very equivocal and hesitating early phase, it was opposed to the construction of a reservoir at the Cow Green site. The 'Teesdale Defence Committee', as the 'ad hoc' co-ordinating body for the objectors to the Bill was called, also used expert witnesses; mostly professional botanists of unchallenged eminence who were prepared to give evidence voluntarily because they were opposed to the scheme.

In addition, however, many scientists acted as partisan supporters of the Teesdale Defence Committee, either as members of the Committee or as advisers to it. Max Walters, a member of the present 'Science and Ethics' Committee, was one such scientist, and he tells the story in the following way.

It was our responsibility to steer the political opposition, to organize public meetings and to plan publicity, especially for the financial Appeal. Here the tensions between objective science and hot politics were most obvious, and most of my colleagues and myself showed signs of strain during the controversy. Some of this was inevitable, because the affair was time-consuming, technically complex and protracted, and had (so far as we were concerned) to be fitted into our normal professional life. Part of the strain, however, was undoubtedly due to the unfamiliarity of the roles we found ourselves playing, and the difficulties which this faced us with, as scientists.

One such difficulty, which arose in an acute form very early in the controversy, concerned certainty about matters of fact. No detailed maps, either of the distribution of the rare species or of the rare communities, existed even in unpublished note form, nor was it possible to obtain quickly such accurate information. Even the statement that, for example, the species *Minuartia stricta* grew nowhere else in Britain, when challenged, might, in scientific honesty, be reduced to 'has not been found anywhere else in Britain'. Even within the area normally assumed to be factual and indisputable by scientists, it is easy to see that legal cross-examination can mount an effective attack.

Similar difficulties arose from statements of prediction and probability. Some predictions are more soundly-based than others, but scientists are trained to be cautious about all predictions. In the Teesdale controversy, they had no difficulty in establishing the 'fact' that some of the rare plants and communities would be destroyed by inundation, if the reservoir was constructed, but the assessment of probable survival adjacent to, but outside, the inundated area caused obvious difficulty. When cross-examined, they could only say it seemed to them highly probable that such rare (and presumably genetically depauperated) species would be unable to adapt to whatever microclimatic and hydrological changes the reservoir would cause. Assessment of probable damage during construction was even more difficult. In these circumstances, scientists were 'unsatisfactory' witnesses for the opponents' cause.

Other problems arose from the assessment of the uniqueness of the site. Scientists for the Teesdale Defence Committee claimed that the threatened communities were unique relict survivors from a late-glacial vegetation type widespread in Britain 10–12,000 years ago. This statement was challenged in two (opposite) ways: on the one hand, they were asked 'How can you be so sure, if you have not accurately surveyed the whole land surface of the Northern Hemisphere?' and on the other the logical point was made that all communities are unique in the sense that no two can be exactly alike. Such attacks disconcerted the scientists, and appeared to them no more than legal casuistry.

Perhaps the most telling difficulties, which strike at the heart of the scientific method, arose in the selection of evidence. Both sides in the controversy selected and used those 'facts' which supported their case, but in the nature of the game the opponents of the reservoir had to bear the main burden of providing and substantiating facts for the promoters to attempt to refute. (This imposed a heavy burden on the Teesdale Defence Committee, for the collection and presentation of data is expensive.) For the scientists, the biased selection of facts and figures was a worrying affair which they did not find easy, in spite of the obvious necessity to play the game according to these rules.

Despite these difficulties, the Teesdale Defence Committee succeeded in mounting a considerable campaign in opposition to the building of the reservoir. In the end, however, they failed; the Bill was passed and the area was flooded. Was the whole effort wasted? Max Walters thinks not. As he put it, 'I believe that substantial gains resulted from the controversy which, it seems, could not have been achieved in any other way.'

What were these gains? One was undoubtedly that public opinion was alerted to the urgency and importance of questions concerning the conservation of nature and natural resources, and people showed by the size of their support for the Teesdale Defence Committee how concerned they were. Of course the support came from a minority, but all 'good causes' begin inevitably as minority concerns, and nature conservation is no exception. The 'Cow Green Reservoir Battle', as the press called it, was undoubtedly one of several factors which, working together, produced Government recognition of the importance of environmental issues (so that we now have a Department of the Environment!) and international governmental support for the UN Stockholm Conference on the Environment in 1972. The fact that an oil crisis and a general recession have obscured our view of many of these problems since does not detract from the achievement, nor remove their urgency.

On the narrower issues of planning of water resources and the siting of reservoirs, the size and effectiveness of the opposition over Teesdale certainly meant that subsequent schemes are much more sensitive to environmental issues, and the authorities concerned are more careful to consult beforehand. Indeed, the whole question of piecemeal exploitation of water resources was sharply to the fore in the Teesdale affair, and regional water authorities would now find such lack of adequate long-term planning very difficult to defend. (Though strictly a separate issue, the drought of 1976 also highlighted the same weakness in the planning of water supplies.)

The stringent conditions accepted by the promoters and attached to the passing of the Act meant that, possibly for the first time, adequate precautions were taken to minimize damage during construction to the adjacent surviving rare vegetation types. Wherever such conditions are appropriate, they are now increasingly made on scientific advice when planning consent is given.

As a result of a Teesdale Research Fund of £100,000 donated by Imperial Chemical Industries (the main beneficiaries of the reservoir scheme), very accurate mapping and monitoring of the surviving communities is being carried out. Ironically, we probably now know more in more detail about this particular small piece of vegetation than almost any other in the world.

Within the Botanical Society of the British Isles (the specialist scientific society primarily responsible for the opposition) the 'esprit de corps' generated by the Teesdale battle has undoubtedly greatly strengthened its corporate concern for the conservation issues. The Society has played a leading part in bringing on to the Statute Book the Conservation of Wild Creatures and Wild Plants Act which finally became law in 1975. When it is realized that the Society was wholly unconcerned with conservation issues until the post-war period – even to the extent of condoning the collection and distribution of dried specimens of nationally rare plants – the size of the revolution in its affairs and its publicity can be appreciated.

The 'Teesdale affair' is trivial when placed beside the life-and-death problems of the medical scientists or the global issues of world food supply and starvation. Nevertheless, one thing may be common to all these issues. It is this. Many scientists, even those working in obviously applied fields, view their studies as neutral, objective, and unconcerned with ethics or politics. As citizens or human beings they make decisions in which they see judgments or even emotions involved, but they somehow cling to the idea that such frailties are inappropriate to their science. This tends to isolate them from the 'real world' of business and politics, and make their pronouncements seem naïve. The respectable reason for behaving like this became very apparent during the Teesdale enquiry: scientists were expected to supply factual evidence relevant to the enquiry, and could only do so effectively if their honesty were unchallenged. But lawyers and politicians in particular, and human beings in general, do not argue like this. We select to suit our case. What then should the scientist do? As a result of his experiences in this enquiry, Max Walters drew the following conclusions. 'I think Teesdale taught me that a *via media* was necessary. I could not in practice say "as a scientist I tell you this, but as a human being I say that". Descent into the arena of politics (or for that matter everyday life) means that some compromise is necessary. You cannot keep your hands scrupulously clean: if you do, many causes dear to your heart will be lost through default, and no one will benefit from your defection. But even if this much is accepted, we are still left with the problem of guidelines. What kinds of compromise do we accept, and under what circumstances? Does the degree of compromise depend on the cause?'

References

AAAS, 'The Integrity of Science', A Report by the AAAS Committee on Science in the Promotion of Human Welfare, *American Scientist*, 53 (1965), pp.174–198.

Bradshaw, M. E., 'The Teesdale Flora', reprinted from J. C. Dewdney (ed.), *Durham County and City with Teesdale*, British Association Handbook (1970).

British Association, *Science and the Media*, Report of a Study Group (1976).

Godwin, H. and Walters, S. M., 'The Scientific Importance of Upper Teesdale', *Proceedings of the Botanical Society of Britain*, 6 (1967), pp.348–351.

Gregory, R., *The Price of Amenity* (1971). The relevant chapter from this book is also reprinted in P. J. Smith (ed.), *The Politics of Physical Resources*, Penguin Books in Association with the Open University Press.

8 The Conservation of Nature

These fragments have I shored against my ruins
T. S. Eliot

Each of the six case studies presented so far explores one particular aspect of science, and the ethical issues which arise from it. This chapter, and the next, will look in more detail at two specific, and very important, ethical issues: conservation, and defence. Both, of course, are huge subjects which merit much fuller treatment than they can receive here. But even a cursory look will help to fill out the picture of the ethical undercurrent in science, and prepare the way for a more detailed discussion of the relationship between the methods, practices, and values of science, and ethics itself.

Behind the Teesdale Affair, discussed in the last chapter, lie all the complex ethical questions of conservation. Why is it important? How can the value placed on conservation be measured against other values?

Dip into any book on nature conservation, and sooner or later you will probably find a reference to the passenger pigeon. During the last century, up until the late 1870s, one of the most imposing wildlife spectacles in the United States was the huge migrating flocks of these birds, flocks that could be over a mile wide and a hundred miles long. Their total population was several billion, far exceeding the total human population of the earth at that time. Yet, by 1898, the species was practically extinct. Though a few survived into the present century, the last of the passenger pigeons, Martha, died in Cincinnati Zoo in 1914. No one set out with the intention of exterminating the passenger pigeon but, through destruction of its habitat and unrestricted hunting, it was exterminated all the same. In less than two generations and without such sophisticated aids as aircraft, chemical defoliants or weedkillers, but with only hand, axe, gun and trap, a flourishing and abundant species was totally eradicated from the face of the earth.

The passenger pigeon is quoted because it provides a spectacular example of the devastating effect that man can have on his environment. But there are many others. Over twenty species of birds have become extinct in the United States alone since the coming of the white man, and the number of endangered species of animals and birds runs into many hundreds (detailed lists are published by IUCN in their Red Data Books). Of course, not all species are in danger. The grey squirrel, the common pigeon, the Norway rat and the water hyacinth are flourishing to an almost embarrassing extent. 'Fewer than ten percent of all mammals are clearly faced with the threat of extinction' (Ehrenfeld, 1972, p.200) but many species are so threatened, and it is very unlikely that all can be saved. Some, like the orangutan, the Siberian tiger or the blue whale, are given widespread publicity and, on occasion, disproportionate amounts of help. The attempted Anglo-Soviet mating of giant pandas may have been a brave attempt, but at times it verged on an expensive farce. Other fascinating or potentially useful species, such as the aye-aye or the addax, receive far less attention, whilst the passing of many insect and plant species goes almost unnoticed. Indeed, some forms of life are eradicated with deliberate vigour. Whatever else the smallpox virus may be, it is also a unique species which man cannot recreate once it has gone.

The gradual extinction of different species of animals and plants, and their replacement by others, is an essential part of evolution. The natural time-scales are so long, however, that the present rate of extinction for probably all species is due, directly or indirectly, to the activities of man. War, the spread of cities, hunting, pollution, the use of chemicals in agriculture and pest control, the introduction of foreign species by man, collecting for zoos and museums, open-cast mining, the building of reservoirs and canals – all these play their part. Sometimes a species is endangered or eradicated almost accidentally, as in the case of the ivory-billed woodpecker, some-times as a direct result of over-exploitation, such as the blue whale, or from a deliberate attempt to reduce the numbers of a supposedly 'dangerous' species, such as the Indian wild dog. Occasionally, it is a result of apparently the best of motives. 'Dozens of other species or strains are now imperilled by the Green Revolution, from the unique strains of African rice to wild fruit trees of the Malaya and Amazon rain forests; most will never even be reported as missing when they go' (Ehrenfeld, 1972, p.53).

It is the accelerating rate at which species are becoming extinct, the real danger of decimating wild species in the next few decades, and the fact that extinction is an irreversible process, which are causing concern amongst scientists, conservationists and others. Not that there has been no response to these dangers. The impor-tance of conservation has been recognized by many Governments, particularly in black Africa, and there is a plethora of national and international bodies dedicated to it, like the International Union for the Conservation of Nature and Natural Resources (IUCN) and the World Wildlife Fund. Occasionally, species can be preserved by careful husbandry in zoos or game parks: the European bison and Pere David's deer are obvious examples. But zoos have only a limited role to play. Inevitably, they can preserve only a relatively small number of animals, and the risks of inbreeding and genetic depauperation are very real, though the extent of these dangers is not fully understood and, according to some experts, may have been exaggerated in the past (see Tudge, 1976). Then there are the problems raised by those species, such as Przewalski's horse, which need a proper herd structure to breed successfully, which can be very expensive to maintain. Other species simply fail to breed successfully in captivity. The orangutans, for example, have a very poor breeding record for animals which have been raised away from their natural environment, whilst the failure of the flightless owl parrot of New Zealand to breed, after many years in captivity, emerged only as a result of postmortem examination; all the specimens were male. New Zealand biologists have failed, so far, to find a female (Tudge, 1976).

Because of this, there is a growing realization that most species need their natural environment if they are to survive successfully. Conservation then becomes a matter of habitats, not merely of species. It implies the protection and preservation of wild species – animals, birds, insects, plants and even bacteria – in their natural surroundings, of complete and essentially self-contained ecosystems, areas of wilderness in which man has no place. Without such areas of wilderness, the diversity of life on earth may be drastically and irreversibly reduced by the activities of man. A species like the ivory-billed woodpecker can only exist as part of a complex ecosystem; it is forced by its unique behaviour and physiology to occupy some inflexible niche in a total system, and any disturbance of that system can wipe out the species. But ecosystems themselves can be very vulnerable. The disruption of the migratory routes of caribou by fencing and pipelines in the Arctic, or the introduction of rabbits to Australia, or water hyacinths to Florida, can put at risk many other species which inhabit the same area. Zoos are not enough; nature reserves and protected areas are required, together with legal powers to control commercial and casual exploitation of threatened species, and, above all, an appreciation through education of the need to protect and preserve the full variety of life in its natural surroundings; to set aside areas of this planet which are for species other than man.

Unfortunately, conservation can be an expensive business. F. E. Warburton has estimated that a single pair of Siberian tigers, with cubs, needs at least five hundred square miles of woodland to meet daily food requirements (Ehrenfeld, 1972, p.289). Assuming that at least one hundred pairs are required to maintain a viable breeding population, then the reserve dedicated to their preservation would need to be the size of England; a lot of land to dedicate to one species. National parks and protected areas need money to set them up, and money to look after them. Moreover conservation prevents resources from being used for the immediate good of man, saying that the interests of the future are more important than those of the present. It is one thing for rich Europeans to set aside nature reserves for their endangered species, but it is something very different to ask the poor, undernourished and overpopulated countries of the world to spend scarce resources, and deprive themselves of potentially useful agricultural land, in the name of conservation. Conservation in Britain, and conservation in India, involve very different restraints. A rich country may put some of nature aside for future use, but if many people are starving then it can seem like callousness.

Most conservation issues involve two types of conflict. Firstly, there is the conflict between the tangible and the intangible. The benefits from any new economic project, whether it involves building a new reservoir, clearing unspoilt forest, or hunting whales, can normally be clearly identified and measured in readily understandable terms: financial savings, increased prosperity, more jobs and so on. On the other hand, opposing values may be vague and diffuse. How does one measure the value of the ivory-billed woodpecker, or a beautiful landscape? How does one set about, as Harvey Brooks put it, 'bringing them into a common intellectual framework with the rest of the analysis' (Tribe et al., 1976, p.116)? It is like trying to decide whether you prefer eating chocolate to watching a football match; without some means of direct comparison, any rational decision becomes impossible. Cost-benefit analysis, which will be discussed in a later chapter, offers some help but, to quote Harvey Brooks again, these decisions 'are fundamentally political in the sense that they ultimately involve competing or conflicting values, and therefore cannot be resolved by purely "rational" (that is, empirical and logical-deductive) means' (ibid., p.115).

The second type of conflict arises when one tries to balance the needs of the present against the needs of the future. What responsibilities, if any, do we have to future generations? And how should we discharge them? Passmore has argued that 'we can reasonably be expected ... to hand on to our immediate posterity a rather better situation than we have ourselves inherited' (Passmore, 1974, p.86). This might seem a very reasonable point of view, but it begs a lot of questions.

What, for example, is our 'immediate posterity'? We have responsibilities to our children, and our grandchildren, and possibly even our great-grandchildren. But what then? Are we responsible for all eternity? Clearly, in some cases, our responsibilities do extend indefinitely: any action whose effects are irreversible implies a responsibility for all time. But in others the degree of responsibility does fade into the future, and where the dividing line comes is by no means clear.

Then there are the problems of what constitutes a 'better situation'. Some have argued that the present rate of consumption of oil and mineral supplies is irresponsible since they will be rapidly exhausted and there will be none left for future generations. But even this argument is not a simple one. Perhaps more oil should be left for the future, but it is equally possible that 'the curtailment of present consumption of oil would impair current productivity and result in less capital of other kinds being left for future generations' (Vickrey, 1953, p.51). Similarly, one might argue that if fewer great apes are used in medical research, in an effort to conserve the species, then our abilities to discover new drugs and cures for diseases will be impaired, with a resultant loss of average human health and life-expectancy in the future. Nor can we be sure that what we regard now as a 'better situation' will be so viewed in the future. 'What we call progress', remarked Havelock Ellis, 'is the exchange of one nuisance for another nuisance.' A cynical remark, perhaps, but one which underlines the relative instability and transient nature of this kind of value judgment.

But it is worse than this. Even if we could decide what kind of life our descendants would like, we do not know how to create it for them. It is true that what we do now determines what happens in the future, but our proven inability to predict the effects of our actions would seem to militate against any assumption of responsibility. Long-term economic and social forecasting is still in its infancy, and its failures are more common than its successes. As Lord Ashby said in the Fawley Lecture in 1975, 'in these turbulent times, even the most percipient expert cannot expect to make credible, let alone accurate, numerical predictions by extrapolating trends' (p.4).

Passmore's statement, then, needs a little tightening up. We have responsibilities to future generations, but those responsibilities are rather better expressed in the following way. We can reasonably be expected to avoid those courses of action which would inevitably deprive posterity of the possibility of enjoying those benefits which we now enjoy. This may seem a rather negative way to put it, but it avoids suspect value judgments as to what posterity will enjoy, and shifts the emphasis of the statement from the future to the present. The objective is not to choose a 'better' future and plan towards it, but to ensure that possibly desirable options are not removed by destruction of natural resources now. Moreover, whether these responsibilities extend for one generation or a million is irrelevant; the next generation can make its own choices.

Seen in this light, conservation becomes a form of insurance. If areas of wilderness are set aside then they will provide reserve supplies of natural species which man can use and enjoy in the future, should he so wish. If such areas are not created, then he will be deprived of even the possibility. It may be expensive to do this, but then so is any form of insurance. How much one is prepared to pay depends upon the value of what is being insured.

Why is nature valuable to man? There are a thousand reasons, which can only be touched on here. In one sense, biological species can be regarded as just another natural resource. They feed us, clothe us and provide many of our goods; they can be used in medical research, and form a branch of study in their own right. Indeed, the discoveries of ethology may have profound implications for psychology, sociology and even politics. The most unlikely species may represent a potentially valuable resource. The addax, for example, might provide an invaluable aid to 'farm' the deserts of North Africa, provided it can be saved from the threat of extinction caused by over-hunting. Even the smallpox virus may need to be saved. 'To attempt totally to wipe out dangerous species of virus and bacteria . . . might encourage the emergence of still more dangerous mutants' (Passmore, 1974, p.102).

Biological resources differ, however, from mineral deposits and other similar resources in a number of very important ways. Whenever copper or iron or any other mineral of this type is used, it is not destroyed and it can, in principle, be recycled (though this may be very expensive). Similarly, alternatives may be available; even if oil is running out, there are plenty of other potential sources of energy – solar power, wave power, geothermal power and the like. The orangutan, on the other hand, is unique and, once used, cannot be recycled. Like Humpty-Dumpty, once it has fallen, it can never be put together again.

Then again, the over-exploitation of copper does not, of itself, greatly affect the supply of iron, or coal, or any other mineral. Biological resources, however, interact in a highly complex way, and even partial removal of one species may ultimately destroy a whole community.

A third difference arises from the diversity of life. Oil from the North Sea is very much like oil from anywhere else, so that if one supply of oil dries up, provided there are ample reserves of oil elsewhere which are still available, no one is too worried. However, all locally autonomous populations of a given species are distinct, in that they differ slightly in their genetic makeup. These differences are very important; for example, one population of a given species may be resistant to certain diseases which would decimate others, and any reduction in the diversity of a given species will lower its resistance to disease. There is no better example of this than the Green Revolution. By careful cross-breeding of the many diverse forms of certain plants, strains have been produced which can greatly increase the yield of a given area of land. These strains are now supplanting traditional varieties, many of which are becoming extinct. 'The great tragedy of the Green Revolution as currently pursued is that it tends to destroy the very diversity that it and the world need to survive and prosper . . . Green Revolution scientists expect to keep one step ahead of crop diseases by continuous breeding programmes; where the breeding stock is to come from in an efficient, monocultured earth they do not say' (Ehrenfeld, 1972, pp.49, 54).

However, the resources of nature do not only enrich the practical or functional aspects of human life; they also give pleasure. Living species and natural wildernesses evoke an aesthetic response, a pleasure in the existence of something beautiful. Even snakes, spiders, and bacteria have their devotees. Not everyone responds to nature in this way, just as there are those who regard public expenditure on works of art as a waste of money. But both the Mona Lisa and the aye-aye give pleasure to many people, both require some measure of protection, and the destruction of either would be regarded by many as a tragedy. Even when a species is of no apparent functional use to man, and its extermination would not appreciably affect any other species, nevertheless its mere existence as a distinct species may give pleasure and lead one to try to preserve it if possible. The ivory-billed woodpecker was, in many

ways, a fairly unimportant species, but its loss, for it is almost
certainly extinct although it is still officially described as one of the
most endangered species in the world (Fisher et al., 1970, p.225),
has been mourned by many; this alone may have been sufficient
reason to attempt to preserve it.

Music, of course, costs almost nothing to preserve, and the preservation of paintings is relatively cheap. Buildings are rather different. The Georgian centre of Bath was both beautiful and unique;
however it was felt by some to be a hindrance to social progress and,
for this reason, parts of it were torn down. Something of beauty has
gone, something of functional utility has replaced it. This kind of
choice is a common one, but the answers are rarely obvious. As
Vickrey has put it: 'The relative place of material progress and
sentimental nostalgia in our scale of values is a real problem' (Vickrey, 1953, p.52). Sentimental nostalgia is not the only argument in
favour of conservation, but it is one of the hardest to relate to other,
more tangible, benefits.

As so often happens, it all boils down to money. Of course we
want to save the Siberian tiger, but can we afford to do so? The
difficulties raised by attempting a cost-effectiveness, or even a
utilitarian, balance of the values and costs of nature conservation are formidable, as has already been stressed. It is possibly for
this reason that many conservationists have based their arguments
on statements of principle rather than utilitarian calculations.
Such statements do not necessarily make individual decisions any
easier, but they at least provide an important counterbalance to the
powerful economic arguments which are used against conservation
measures.

There is no shortage of such statements in the literature, which is
vast and, for the most part, clearly committed to the cause of
conservation. Unfortunately, they are not all very sensible. As so
often happens, the zealous reformer, impatient with his fellow men,
tends to overstate his case. Conservation can be justified purely in
terms of human interests: that it is wrong unnecessarily to destroy
nature and natural resources, thus preventing future man from
enjoying and using them. Unnecessary pollution, over-exploitation,
casual disregard of the value of such resources, or sheer wanton
destruction, can only be justified and said to be necessary if the
benefits of using resources in this way far outweigh the benefits of
preserving them. Otherwise, they are manifestations of human
greed or arrogance, exploiting resources for inadequate short-term
gains, wasting them in disregard of their true worth or, in many

cases, destroying them for no real purpose, which is simply an act of vandalism. Failure to consider the values of conservation, or failure to weigh all the benefits, costs and risks involved before beginning a particular course of action, is, for these reasons, morally reprehensible. Of course, one can argue forever about what is necessary and what is unnecessary; all one can say is that there must always be an element of proportion.

Those conservationists who try to go beyond this position, and argue that our responsibilities are not just to present or future man but are to nature itself, do so for a variety of reasons. For some, it is an expression of a kind of nature mysticism which, though it is not a part of the mainstream of Western culture, nevertheless has a long lineage through writers like Wordsworth, Emerson and Thoreau (Passmore, 1974, p.28), and which leads naturally to the idea of responsibilities to nature, almost as a matter of faith. Others, without actually venerating nature, argue for a growing awareness that man is a part of nature, a view which modern science can seem to reinforce, so that his system of values must extend to the whole rather than a part. Here again, it is possible to trace this view through history and such writers as Victor Hugo (ibid., p.3), but its recent importance within the conservation movement is due more to writers such as Aldo Leopold. 'That land is a community is the basic concept of ecology, but that land is to be loved and respected is an extension of ethics. . . . We abuse land because we regard it as a commodity belonging to us. When we see land as a community to which we belong, we may begin to lose it with love and respect' (Leopold, 1970, *Foreword*).

If the adoption of such views is regarded as an act of faith, then there is little room for discussion: they can either be accepted or rejected. However, most writers who advocate such views tend to use two, interrelated, arguments. Firstly, they argue that this extension of ethics is the next step in a natural progression. 'The first ethics' according to Leopold 'dealt with the relation between individuals; the Mosaic Decalogue is an example. Later accretions dealt with the relation between the individual and society. . . . There is as yet no ethic dealing with man's relationship to land and to the animals and plants which grow upon it. Land, like Odysseus' slavegirls, is still property . . . entailing privileges but not obligations. The extension of ethics to this third element in human environment is . . . an evolutionary possibility and an ecological necessity. It is the third step in a sequence' (ibid., pp.238, 239). The main difficulty with such arguments is that they blur the fact that it is not just the third step but a major conceptual leap. The implication is that this new ethic would not be based exclusively on human considerations and values (or, for a Christian, on our duties and responsibilities to God), on which the whole of our present system of ethics is based, but on the attribution of values to nature which are wholly independent of human considerations: that nature has rights independent of man. The very words used point out the difficulty. It would be perfectly normal to talk of cherishing land, but to 'love and respect' land implies much more. The change of thinking is too great to call this new ethic an 'evolutionary possibility'. A 'revolutionary possibility' it may be, but then it has more of the flavour of an act of faith than a result of logical thinking.

The second argument for the new ethic is because, in Leopold's words, it is 'an ecological necessity'. Here, the basic premise is that western traditions are inadequate to deal with man's relationship to his environment. The tradition which the conservationists find is 'that nature exists primarily as a resource rather than as something to be contemplated with enjoyment, that man has the *right* to use it as he will, that it is not sacred, that man's relationships with it are not governed by moral principles' (Passmore, 1974, p.20). It is a tradition which is often summarized in the phrase 'the dominion of man' (see, for example, Black, 1970), but it is a tradition which is widely misunderstood. The historian, Lynn White, for example, has tried to argue that Judaeo/Christian beliefs are actually responsible for the 'rape' of nature. 'Both our present science and our present technology are so tinctured with orthodox Christian arrogance toward nature that no solution for our ecologic crisis can be expected from them alone. Since the roots of our trouble are so largely religious, the remedy must also be essentially religious, whether we call it that or not. We must rethink and refeel our nature and destiny' (White, 1967, p.1207). Not only does this completely ignore elements such as the Franciscan attitude to nature, but it misunderstands the meaning which theology attaches to the concept of dominion. Dominion is not absolute in the Judaeo/Christian tradition, but qualified. Men are to rule in righteousness (or justice) and mercy as the God did in whose name they ruled and under whom they exercised authority (Dunstan, 1974). Man is God's steward for nature, not its absolute ruler.

Leopold's arguments are echoed in Stone's book on the legal rights of natural objects (Stone, 1974). Children, women, subject people formerly enslaved, trusts, corporations, institutions, municipalities – all these have been given rights in law because, Stone argues, this is the simplest way to protect the underlying human interests involved, and his thesis is that nature should be given rights for the same reason. His ideas have received considerable support (see, for example, Ashby, 1976) but are open to the same criticisms as those of Leopold. It is not the only method available to protect the environment: the defence of the Teesdale flora did not fail because the flowers lacked legal rights, nor is it likely that it would have succeeded if they did have such rights. Moreover, ascribing rights to natural objects is fraught with ethical difficulties. Passmore, in his carefully reasoned essay on the subject, argues that, whilst men form a community with plants and animals in the sense that they are ecologically interdependent, they do not recognize mutual obligations, nor do they have common interests so that 'in the only sense in which belonging to a community generates ethical obligation, they do not belong to the same community'. His conclusion is that 'the idea of "rights" is simply not applicable to what is non-human' (Passmore, 1974, p.116). He admits to being 'troubled by the apparent dogmatism of my observations', but it is clear that extending the idea of rights to natural objects is not just the extension of ethics but a complete overhaul of the system. Such an overhaul may be possible in principle, but it is not obviously necessary nor, in view of the difficulties involved, is it obviously desirable. In practice it matters little whether one regards one's responsibilities as being to God, to future mankind or to nature itself; in each case a value is placed on conservation, an injunction not to over-exploit or wantonly destroy natural resources, tempered by the needs of present man. (Those who try to make nature conservation an absolute principle look very foolish when one points out that the cause of most species would best be served by the extinction of man!) Talking in terms of 'loving' or 'refeeling' nature may help to sway public opinion, but it has little effect on the essentially practical arguments for and against conservation.

How do the issues of conservation affect the practising scientist? Obviously, he cannot ignore them. Whether he is collecting plants, experimenting with new drugs, devising new pesticides, fishing coelocanth from the sea for scientific study, breeding more productive forms of wheat, helping to produce new and more efficient industrial processes – in every case he is making implicit value judgments on the relative importance of conservation, though he may not realize it. Clearly, every scientist must be aware of the issues involved and make his decisions from a position of knowledge, not ignorance; failure to do so is to deny his responsibilities as a human being.

Science also has a duty to discover and make known the facts, as far as possible. We know a great deal about the larger mammals and birds – their population sizes, habitat needs, histories and so on – and can use this information to make sensible decisions as to their conservation. But many species, particularly the lower animals, plants, and insects, are so little studied that they may become extinct before they have even been identified. Like throwing away unopened letters, this could well be a very foolish course of action.

Whether the scientist has a duty to go beyond the facts and press for conservation is rather different. There are those, including many scientists, who would argue that the job of the scientist is merely to provide a dictionary of facts; it is society's job to string those facts together into coherent sentences of social policy. But it is almost impossible to present facts in a wholly objective way, and the decision facing the scientist is whether to try to present the facts as objectively as possible, whilst at the same time revealing his personal prejudices, or whether to sally forth in the name of conservation, which does not lack its crusading zealots within the scientific community. Most scientists probably believe in the principle of conservation; the extent to which they are prepared to champion it, or to compromise in favour of other interests, must depend on personal beliefs. Either way, as the Teesdale Affair shows, the scientist cannot avoid ethics.

Conservation is one of the fundamental problems of the present century; it is also one of the most difficult. Manipulating his environment on a global scale has brought many benefits to man, but it has also demonstrated his capacity to destroy his environment. Conservation may bring some insurance against this destruction, but it will also reduce the level of immediate benefits. To western man, this merely implies a tightening of belts; to much of the third world it may imply poverty, disease, starvation and death. These are not problems which can be avoided; we have a duty to try to find a solution which will safeguard the interests of the future without compromising those of the present. It is not beyond the wit of man to do so.

References

Ashby, Lord, 'A Second Look at Doom', Twenty-first Fawley Lecture, University of Southampton (1975).

Ashby, Lord, 'Towards an Environmental Ethic', *Nature*, 262 (1976), pp.84–85.

Black, J., *The Dominion of Man: the Search for Ecological Responsibility*, Edinburgh University Press (1970).

Dunstan, G. R., 'Man's Dominion and the Nature of Man', in *Science and Absolute Values*, International Cultural Foundation Inc., New York (1974).

Ehrenfeld, D. W., *Conserving Life on Earth*, Oxford University Press, New York (1972).

Fisher, J. and HRH The Duke of Edinburgh, *Wildlife Crisis*, Hamish Hamilton and World Wildlife Fund (1970).

Leopold, A., *A Sand County Almanac*, Balantine Books, New York (1970), first published 1949.

Passmore, J., *Man's Responsibility for Nature*, Duckworth (1974).

Stone, C. D., *Should Trees have Standing?*, Kaufmann (1974).

Tribe, L. H., Schelling, C. S. and Voss, J. (eds.), *When Values Conflict*, Ballinger, Cambridge Mass. (1976), especially H. Brooks, 'Environmental Decision Making: Analysis and Values', pp.115–135.

Tudge, C., 'Last Animals at the Zoo', *New Scientist*, 71 (1976), pp.134–136.

Vickrey, W. S., 'An Exchange of Questions Between Economics and Philosophy' (1953), reprinted in E. S. Phelps (ed.), *Economic Justice*, Penguin Books (1973).

White, L. Jr, 'The Historical Roots of Our Ecologic Crisis,' *Science*, 155 (1967), pp.1203–1207.

9 Defence and the Scientist

> When it's a question of peace one must talk to the Devil himself.
>
> *Herriot*

Living species cannot exist in isolation, but only as a part of a complex, interdependent ecosystem. This is why there is an increasing recognition that a successful policy of conservation must be about areas of land, not merely individual species. Moreover, these areas of land must be an adequate size. Land can only support a certain density of a given species, above which there will be insufficient food and the species as a whole will suffer accordingly. Overpopulation is just as much a problem to blackbirds, lions and oak trees as it is to man.

Nature avoids overcrowding in a variety of ways, some of which are quite subtle. One common method, particularly amongst the higher vertebrates, is for each individual or group of individuals of that species to acquire a territory, an area of land which other individuals of the same species are not allowed to enter. Sometimes a territory will be 'owned' by a single pair of animals, with offspring, sometimes by a somewhat larger group of animals. In each case, however, the effect is the same: to ensure that there are never more animals of that species in a given area than it can comfortably support. Indeed, 'obtaining or not obtaining a territory can be the most crucial single item among the traits and achievements that make up an individual's fitness' (Wynne-Edwards, 1974, p.9). Failure to do so can be disastrous; each year, for example, more than sixty per cent of the new generation of red grouse on the Scottish moors fail to obtain a territory and, for this reason, die during the winter. 'Yet the contest for territories is often won or lost through some ritualised encounter. . . . Enduring decisions are reached without either participant having to inflict serious physical injury on the other' (ibid., pp.9, 11).

Whether or not such observations on other species have any relevance to human behaviour is one of the more contentious areas of modern science. But it is difficult to ignore the apparent similarities between territorial behaviour in human beings and in other species. Nor do the threat and counter-threat of so much of international politics look so very different from the ritualized encounters and aggressive posturing of other animals. There is, of course, an important difference. 'There are species like the large herbivorous mammals, many of them gregarious, which already possess exceedingly lethal weapons in the shape of horns and antlers, and have unlimited opportunities to take a rival unawares, and, in a moment, inflict on him a mortal wound; but they are totally inhibited from doing so' (ibid., p.11). War, and the killing of an animal by a member of its own species, are extremely uncommon phenomena in the animal world. The same, alas, is not true for man. 'War' according to Bismarck 'is, properly speaking, the natural condition of humanity.'

Bismarck overstates the case, of course. But defence and national security, if not war itself, are a central preoccupation for most Governments, and ones which inevitably raise considerable ethical problems. Pure defence, provided that the dangers which it seeks to overcome are real and can be clearly demonstrated, is relatively easy to justify. But defence is not so far removed from offence. The threat of attack is often the best means of defence, but then there is always the possibility that threats may become realities. Defence inevitably involves recognizing the possibility of war, and cannot avoid raising all the ethical difficulties which war brings in its wake.

Should a scientist become involved in research for defence purposes? If so, what conditions and restraints, if any, should be placed on him in carrying out such research? The answers to these questions depend partly on the attitude of the country to defence, partly on the means by which that country takes the relevant decisions, and partly on the conscience of the individual scientist.

In this country, the broad decisions about the commitment of the nation to defence, and the proportion of the nation's resources to be directed to defence, are taken openly and democratically (though our elected representatives are entrusted with most specific decisions, since they must reserve secret information to themselves). They represent a broad consensual agreement, and it is the duty of the citizen, unless he has an objection of conscience, of which more later, to work in accordance with these decisions. Moreover, since defence is a highly technological affair, society has the need and, presumably, the right to ask scientists to contribute to the defence effort, and some measure of obligation is placed on at least some scientists to undertake the appropriate research.

The extent of those obligations may be a matter of dispute. There is a continuous debate about the extent to which scientific research should be clearly related to the needs of the community, which is explored elsewhere in this report. But if one accepts that the scientific community cannot enjoy a unique position of privilege within society, that science cannot be just the pursuit of pure knowledge for its own sake, unfettered by social constraints, then it is hard to avoid the conclusion that the scientific community has some obligations to use its special skills to ensure the success of democratically-agreed social policies, including defence.

These obligations, however, cannot be absolute. The idea that society should be able to force a scientist, against his will, to undertake defence research is abhorrent to us. If a scientist feels a moral obligation to refrain from active involvement in scientific research for defence purposes, then his conscience should be respected. He may have one of a variety of reasons. He may take an absolute stand against participation in war and in the methods of war on grounds of religious commitment or humanist philosophy. He may take a limited stand against particular forms of warfare as being unacceptably barbarous or destructive, or as being banned by international law. He may conscientiously hold that the preparation of armaments is aimed against some order of society elsewhere in the world which he holds to be ethically desirable. He may have no absolute ethical objection to military methods, but feel that in a needy world a disproportionate amount of energy and resources is being directed to defence. All these are justifiable ethical views, and the scientist who holds them should not be asked to participate in defence research in opposition to his views and his conscience.

Freedom of conscience, however, is not the same thing as freedom from responsibilities. The scientist who contracts out of defence research cannot deny all his responsibilities and ignore the fact that such research is nevertheless carried out. For one thing, there is no clear dividing line between defence research and any other research. Improved radar techniques, or escape-mechanisms in submarines, or even improvements in surgery, benefit society both in defence and in many other ways. In one sense, science is an indivisible whole so that a scientist cannot avoid making some, albeit indirect, contribution to defence. Space research is an obvious example. Even if a participating scientist is apparently concerned only with the scientific exploration of space, he is nevertheless contributing to an overall effort which has profound military implications and uses, as was argued in chapter 6.

But there are degrees of responsibility. Dr Alex Wood, pacifist and teacher of physics at the Cavendish, used to say: 'I distinguish, I hope rightly, between my responsibility for the emergence of the atom-bomb, because I teach the science which enabled it to emerge, and because I taught many of those responsible for its development, and because as a citizen I have failed to persuade the majority of my fellow-citizens of the wrongfulness of war – and the responsibility of those who deliberately went into the atom-bomb project.' To this one can add that it is important to distinguish between the responsibility of the scientist for the use made of his expertise, which he has freely offered, by military authorities, and his inescapable involvement in that use. If, for example, a scientist, through his teaching or his research, has made a contribution to military methods, then he is clearly inescapably involved in the development and potential use of those methods. But if he could not have been expected to foresee that his actions would have military consequences, then it is unfair to attribute any degree of responsibility to him for those methods. One can only be responsible for those consequences of one's actions which one can reasonably have been expected to predict; no one can be held responsible for the unforeseeable. In most cases, however, the scientist will bear some measure of responsibility, though it may be shared by others. If a society develops a missile system, then the greatest burden lies with those who are given the power by society to take decisions as to whether or not to give the order to press the button. But, even though one cannot remove the major burden of responsibility from the man at the top, one cannot isolate him and make him a scapegoat. Some of the total responsibility is shared by the engineer who builds the missile, or the scientist who perfects its aerodynamics or tracking system and, ultimately, by every member of the society which sanctions its development.

There is another sense in which a scientist cannot avoid his responsibilities. Those who, for whatever reasons, deliberately contract out of an obligation undertaken or approved by the majority of their fellow-citizens in a democratic state, have a moral obligation to spend a significant part of their time and energy in seeking to change the situation to which they take exception. This was a point strongly made by Dr Wood. Sir Hermann Bondi has said: 'Basically it seems to me that a citizen either has to agree or to disagree with his Government's defence policy. If he disagrees then it is his duty to make this plain, to think through the alternative he proposes, and to attempt to gather support for his views in the hope of changing the policy. If on the other hand he is in broad agreement, then one would expect him to co-operate and, when appropriate, to assist in carrying through this policy. Either of these courses of action strikes me as entirely ethical and honourable, whereas the attitude of not wishing to have anything to do with defence because it is "dirty" while doing nothing to change the policy strikes me as less than honourable.'

Broad generalizations, like those given above, may help to set the issues involved in perspective, but they do not solve all problems. The majority of scientists, for example, would countenance neither a total commitment to military research nor a complete ban on it. The problem then becomes a matter of proportion, of deciding what is allowable and what is not. Sir Hermann Bondi has put the difficulty succinctly: 'The question naturally arises whether over and above the question of defence policy there are limits beyond which the citizen scientist should not go. This is something very difficult to discuss in the abstract: coming down to particular cases is not necessarily illuminating.' The uncertain precedent of the Nuremberg trials suggests a distinction in the minds of the international community between war and war crimes: that there are points when a halt must be called. For Christians the traditional teaching about the just war has sometimes been seen as offering guidance, but contemporary debate suggests that its precise application to present problems is not easy to determine. But there is always a danger, especially in a war situation, of the sliding of moral standards, and it is vital to remain aware of this danger and to encourage continual discussion of the issues involved.

One obvious topic which needs this kind of careful debate is biological warfare. It is easy to argue that biological warfare is a bad thing. It is indiscriminate and particularly unpleasant in its method of attack and, of perhaps greater importance, its effects may well be irreversible and continue to cause fresh human misery and death long after a truce has been made and peace re-established. It is for these reasons that active pursuit of international control over biological weapons is so important, since this can act as a counter-balance to their development by particular countries. But if there is a threat that other nations or groups may use biological warfare against one's country, it is at least necessary to prosecute biological research for the protection of the population through inoculation or identifying agents in water supplies. It is arguable that the possibility of counter-attack by the use of the same weapons is an important deterrent to others, but here there is a shift from pure defence to the threat of attack, and the ethical difficulties increase. There is a real risk that this kind of research, even if it is undertaken in the hope of averting biological warfare, helps to lower the psychological barrier against the use of such techniques. Moreover, it creates the means by which a future government could attack other nations with biological weapons if it wished to do so. There is also the danger that the successful development of biological weapons by one country may make another country feel obliged to refine its own methods in order to keep one step ahead, with spiralling escalation as the almost inevitable result. Then there are all the dangers of accidents in laboratories where new methods of biological warfare are being developed, of leakages into the population at large, or of terrorist raids on such laboratories, resulting in the deliberate release of extremely harmful substances. These dangers cannot be ignored; some of the recent industrial accidents, such as the release of considerable quantities of dioxin near Milan in 1976, have graphically illustrated the risks involved, and the fact that no enterprise of this sort can be absolutely free of risk. But, having said all that, we must also recognize that refusal to undertake such research may leave this country potentially defenceless, and involve simply closing one's mind to a painful possibility which one would prefer not to have to contemplate.

Similar remarks can be made for other military techniques. The unpleasant, indiscriminate and irreversible effects of their use, the arguable need for deterrents but the risk that they will then become offensive weapons, the risks of escalation, or of accidents – all these problems apply just as well to nuclear weapons, for example, as they do to biological weapons. But with nuclear weapons, additional problems are posed by the desirability of nuclear power. Faced with dwindling energy resources, many people advocate the widespread development of nuclear power as a possible solution to future energy supplies. But the spread of nuclear power stations and expertise to new countries inevitably makes it easier for such countries to develop nuclear weapons and increases the risk of their eventual use.

Even more difficult are the problems raised by the use and development of psychological techniques. Most traditional methods of warfare involve physical violation of the body, where the damage done can clearly be seen. Modern psychological techniques however, such as brain-washing, involve violation of the personality, where it is less clear what damage is done. It is this difficulty in assessing the extent of the damage which makes the ethical issues raised by psychological techniques so difficult to clarify. There is a wide spectrum of techniques by which human beings seek to influence one another, ranging from friendly argument and persuasion through advertising, propaganda, to subliminal advertising and, ultimately, to brain-washing itself. At some point along this continuous spectrum, the basic ethical principle of respect for human personality demands that a line should be drawn and a limit placed on those techniques which may be used. But where that line should be drawn is a difficult problem. This applies particularly to the methods used to extract information from prisoners of war, and one might suggest that scientists have a duty to try to discover methods of extracting such information which do not rely on either physical torture or unacceptable mental manipulation, but which lie on the acceptable side of the dividing line.

A second type of difficulty arises from the degree of secrecy which surrounds much defence research. Not that defence work is unique in this. There may well be an injunction against the open publication of newly discovered industrial processes developed through the financial support of a particular firm. Or, to take a perhaps less ambiguous example, there are proper ethical restraints on the open publication of the research results achieved through the Home Office Research Fellowships in Criminology in a way which would reveal the identities of individuals involved. In the case of defence, the arguments for and against secrecy are widely debated, but it is hard to avoid the conclusion that technical details should be kept secret if there is a real risk that their open publication would pose a real threat to national security.

Secrecy, however, is very unwelcome to scientists, many of whom, as was pointed out in the chapter on 'Science in the Open', feel strongly that there should be no restrictions on a scientist's freedom to publish his results or, at least, that these restrictions should be kept to an absolute minimum. This raises two interrelated questions. When, and under what conditions, should a scientist take an oath of secrecy and undertake classified research? And, having taken such an oath, are there any circumstances when he would be justified in breaking it?

Take the second question first. If a scientist has, freely and knowingly, taken an oath of secrecy, then he has an obligation to maintain that oath. If he does not wish to be bound by it, then he should not take it in the first place. If he accepts employment in a defence field he automatically accepts the constraints and obligations which go with it, including the obligation to secrecy. Even if he subsequently leaves defence work, he may recover all the freedoms of the ordinary citizen to oppose Government policy, but he cannot claim a special status through knowing official secrets and divulging them. The obligation to secrecy is not ended when his employment ceases.

This does not mean that the scientist who enters defence work 'sells his soul to the Government'. He retains, and must retain, certain basic freedoms. Suppose, for example, a scientist, who is already involved in defence work, becomes conscientiously alienated from it. He may have discovered, through direct involvement, that he is opposed to this type of work. The work itself may have developed along lines which he can no longer accept. Under such circumstances, he must be able to withdraw, with honour, from the work involved. Not only should he be able to resign, but he should be able to do so in a way which makes it absolutely clear to any future employer that his reasons for resigning were wholly honourable, without necessarily divulging the reasons for his resignation. Such a system can work: there was an excellent example when in 1971 a senior civil servant asked for, and was granted, a transfer because he felt that the Government had put a unilateral and untenable interpretation on the 1926 Geneva Protocol on chemical methods of warfare.

This, however, does not solve the problem of secrecy. Are there any circumstances when a scientist may justifiably break his oath of secrecy? Suppose, for example, he passionately believes that keeping certain information secret is against the national interest. Should he then divulge it? There is no easy answer, no convenient generalization to cover such cases. On the one hand, one can argue that no individual, having taken an oath to abide by decisions which have been reached by a majority through the democratic process, can then have the right to break that oath. But it is possible to envisage situations where the moral decay amongst those who are privy to secret information becomes so dangerous that the breaking of an oath by a scientist in full knowledge of his personal responsibility for the consequences of his action, is the lesser of two evils. How one assesses when that point has been reached is, however, a different matter. Asking the question will at least raise our level of ethical awareness, but it does not make individual decisions any easier to take.

There remains the first of the two questions posed above. When should a scientist take an oath of secrecy? When is it justifiable to prevent the publication of results, and when is it unjustifiable? Obviously, decisions on particular cases must be left to the individual scientists, politicians and military officials involved. But it is possible to offer some general comments. Firstly, secrecy should be avoided if possible, for the reasons given in the chapter on 'Science in the Open'. The open publication of scientific results is an essential stage in the assessment of their reliability; results in defence research which are kept secret may, for this reason, be less reliable than other results and hence create problems. The Starfish and West Ford experiments discussed in chapter 6 amply illustrate this point. The principle of secrecy, once established, has a nasty habit of spreading into areas where it is wholly unacceptable, and some counterbalancing principle is essential to correct this trend.

Secondly, classified research should, as far as possible, be carried out in institutions specifically for that purpose. Universities are not the appropriate place for it, since one of the basic tenets of university research is that it should be the free and open pursuit and communication of knowledge. There are, of course, exceptions. Some industrial research is undertaken at universities on the understanding that its results would not be published until the sponsor has had the opportunity to patent it, should he so wish. In such cases, however, the purpose of these restrictions is clearly defined in the contractual agreement, and there is normally an agreed period of time after which these restrictions cease and the scientist may freely publish his results. For research sponsored by defence departments, however, the situation is more ambiguous. If the research is classified, then the situation is at least straightforward: the scientist must 'sign the Official Secrets Act' and is wholly prevented from making his research public without the consent of the military authorities. But the amount of classified research in universities in this country is small (Fairbairns, 1974/2, p.5) and most of the research falls into a grey area: neither classified nor wholly open. Many Ministry of Defence contracts carry the clause: 'Results shall not be published without the authority's consent. Whilst it is the intention to publish freely in accordance with scientific practice, the authority may, after consultation with the university, delay publication if considered necessary in the national interest' (quoted in Fairbairns, 1974/2, p.5). On the face of it, this seems fair enough, but the wording is sufficiently vague that it can, if interpreted restrictively, effectively classify such research even though it is officially unclassified. One way round this might be to suggest that the purpose of any restrictions on publication was clearly defined in such contracts, and that these restrictions applied only in fulfilment of that specified purpose.

The decision whether or not to undertake defence research can be a very difficult one for many a scientist. Not only must he carefully consider the relationship between his role as a citizen and his role as a scientist, and the demands which society makes of him by virtue of his specialized knowledge and expertise, but, if he undertakes defence research, he may have to agree to some limitation to his normal scientific freedoms. The problems, at least, are clear; how he answers those problems must remain a matter for the conscience of the individual.

References

Fairbairns, Z., 'War Research at British Universities', *New Scientist*, 63 (1974/1), pp.312–315.

Fairbairns, Z., *Study War No More*, Campaign for Nuclear Disarmament (1974/2).

Wynne-Edwards, V. C., 'Society Versus the Individual in Animal Evolution', paper given at the Annual Meeting of the British Association, Stirling (1974).

10 A Scientific Education

The aim of education is the knowledge not of facts but of
values.

W. R. Inge

If some of the threads which run through the previous chapters are
drawn together, a fairly clear picture of the role which many scientists think they should play within society begins to emerge. First
and foremost, they are the bringers of scientific knowledge. Scientists, some would argue, have a fundamental duty to lay bare the
facts, to 'speak out and shame the devil', so that the scientist simply
becomes the instrument by which society as a whole discovers the
truth. Moreover, he must speak not only truths, but the whole truth
and nothing but the truth, and failure to do so is morally reprehensible. Because science itself, they would argue, is capable of total
objectivity, the scientist too must seek to be wholly objective and
should not, in his role as professional adviser to society, permit
himself any statements which are not wholly grounded in scientific
fact. If a scientist publicly champions a particular social or political
policy, which is related to his scientific knowledge, for reasons
which are even tinged with value judgments, however faintly, and
this policy is subsequently proved to have been mistaken, then the
degree of censure he receives from his colleagues may be far greater
than an ordinary citizen would receive.

These attitudes are not universal amongst the scientific community, but they have received widespread support, particularly from
scientific institutions. They are not wrong, but, because they are
presented as absolute commandments for scientists, they are somewhat simple-minded. Science itself is not free of value judgments:
acceptance of the vast construct of mutually compatible statements
which comprises scientific knowledge as 'truth' involves very deep
problems of what one means by 'truth'. Nor is this merely a matter
for philosophical pedantry; the identification of socially defined
terms like intelligence with a quantifiable concept such as IQ rests
ultimately on one's conception of scientific knowledge. Fact and
value are distinct concepts, but it is not always possible to separate
the two. Some of the basic techniques of science – selection of
evidence, generalization from particular examples, acceptance of a

theory for its simplicity and elegance or for its proven predictive power – are similarly dependent on value judgments. Total objectivity is a goal which scientists might strive towards, but it can never be reached. To quote Bentley Glass, 'As long as science is a human activity, carried on by individual men and by groups of men, it must at bottom remain inescapably subjective' (Glass, 1966, p.77).

For this reason, the scientist who steps into the social arena believing that he is capable of total objectivity is mistaken. Indeed, he may prove a positive menace since there is a basic flaw in his powers of judgment. Scientists are still human beings, however much they may disclaim the fact, and not just fact-producing robots. Nor is it possible to separate the scientist from the human being. As Max Walters put it 'I could not in practice say "as a scientist I tell you this, but as a human being I say that"' (though even he speaks of a 'descent into the arena of politics', as if social values not only tinge the scientist's behaviour but taint it as well).

Why do so many scientists believe that science is independent of ethical values and decisions? And why are they so shy of the political arena where ethics are so clearly important? These are not easy questions to answer, but it is possible to pick out some important clues by considering the ways in which scientists are taught, and how they approach their craft: choosing their research project, isolating the problems to be solved, quantifying the variables involved and so on. What, in short, makes a scientist?

Education is obviously the place to start. A scientific training is no more exclusively concerned with facts than an education in history is concerned only with dates. The aim of every teacher in any subject is to guide and influence his students, not to act as a walking dictionary. For this reason, the science student, like any other, emerges from his training with certain attitudes which his teachers have instilled into him, and every teacher has a special moral responsibility towards his students because of his influence over them.

Inevitably, the educational process tends to be self-perpetuating. A good teacher will produce good teachers, a bad teacher will produce teachers who are not so good. Values, too, tend to be infectious. A university professor who concentrates solely on his research students will tend to produce a sense that ordinary people don't matter. A university professor who concentrates solely on his personal research will tend to produce a sense that even his research students don't matter. This also applies to qualities shown in human relationships. A student who admires a scholar's intellectual qualities may, for good or bad, imitate his moral qualities, and the lowering of the age of independence to 18 does not remove the special moral responsibility which a teacher has towards his students. High quality research, which remains the main criterion by which university scientists are assessed, is in principle fully compatible with high quality teaching, as may be seen in a number of great scholar-teachers such as Faraday, but the reality does not always live up to the ideal.

Education involves the initiation of the student into a tradition which, initially, must be taken on trust. Whilst true teaching is a two-way process, which must always include the hope of learning something new from the student, the teacher must inevitably begin from a position of authority, and the student must accept this authority, at least at the start of his initiation. This is true even in science, where the traditions are accepted only because they are empirically based and not merely because they are traditions. The danger in teaching science is that, if too much reliance is placed on authority, particularly during the early years, then the later transition to a rounded appreciation of the empirical basis of those traditions may not be made completely, leaving certain aspects accepted as dogma.

These are matters which concern all teachers, but they serve to emphasize that the teaching of science is a human activity, and that teachers have human frailties and ethical constraints upon them. The main point of interest here is the extent to which science, and the way it is taught, leads to certain attitudes to facts and values, and where and in what way it leads to the attribution of authority. If science graduates differ in their approach to values from graduates in other disciplines, does this reflect the type of person who is attracted to science as a career, or does some of this arise from a scientific education?

Herbert Spencer, in his once famous 'Education', claimed that science was best for both intellectual and moral discipline. He suggested that the learning of languages led pupils to bow to the authority of dictionary or teacher, whereas in a scientific education nothing is taken on authority alone, but there is a constant appeal to individual reason. Spencer was defending the biological sciences from the charge of corrupting the young with evolutionary heresy, and also from the charge (levelled by the physical scientists, who jeered at Darwin's use, more than 800 times, of 'we must suppose') of trusting solely to reason than to experimental facts. However, it is precisely the balance between observation and reason, between experiment and theory, which creates the scientific attitude.

Consider experiment first. Many have argued that one quality which is inculcated by a scientific education is a humble receptivity of whatever nature reveals. Thus, the Puritan, John Webster, believed that laboratory experience for young people would prevent them from 'growing proud with the brood of their own brains', whilst Adam Sedgwick at Cambridge advocated a scientific education on the grounds that 'the study of the higher sciences is well suited to keep down a spirit of arrogance and intellectual pride'. Certainly, the discipline which appeal to experimental proof of reason provides is a hard one, but that is not to say that it leads to humble receptivity. Fact is a treacherous concept in science, for many 'facts' contain within them a degree of reasoned interpretation, and one man's fact may well be another's opinion. Many scientific controversies, such as the recently reported observations of gravitational waves, revolve around just such subtleties of interpretation.

These difficulties become even more pronounced when one turns to theory. Until quite recently, General Relativity was widely accepted as 'truth' despite a lack of experimental verification. Whilst the number of experiments which have confirmed its predictions has increased considerably since 1970, no such experimental tests can ever wholly confirm its validity over all potential rival theories and its acceptance will continue to be based, in part, on aesthetic and philosophical considerations (MacCallum, 1976). Scientific truth, in such cases, owes at least as much to elegance and simplicity as to observable fact. This is even more pointed when one assesses theories of the origin of the universe; indeed, whether it is sensible to extend the concept of truth to any theory concerning a single, unique, historical and unobservable event involving the totality of matter is open to debate.

Humble receptivity, then, is only one side of the scientific coin. On the other side lies the authority which is ascribed to scientific facts, expressed in the language of mathematics. Any field of scholarship relies for its authority on the facts and theories which it can derive by reasoned argument from the observable data; science is special only in the degree of objectivity which it assigns to its derived knowledge, and in its technical success in predicting the future from that knowledge. Rigour in analysis, and elegance of approach, are widely regarded as commendable qualities in science as in any other discipline. This finds reflection in many of the snobberies current amongst undergraduates: chemists will respect physicists and mathematicians but look down on engineers, whilst sociologists are widely regarded as the lowest of the low. These snobberies have two main sources. The first is a widespread feeling that pure science is somehow more noble than applied science, an attitude whose origins have been lucidly exposed by Sir Peter Medawar (Medawar, 1969). The second, however, has nothing to do with the social usefulness of a particular branch of science, but arises instead from the nature of the scientific craft. In most disciplines, the criteria by which the success of a particular piece of science is judged are those which were originally developed in physics. They were subsequently applied, with much success, to other disciplines, first to chemistry, and then to the biological sciences but, for example, chemistry has still not achieved the same success (as judged by these criteria) as physics, nor are the biological sciences yet the equal of chemistry, differences which can produce a ranking order in the minds of students. Behind these attitudes, however, lies a very dangerous assumption: that the methods and criteria of adequacy which have proved so successful in physics are automatically appropriate to other disciplines. Many of the failures of economics, in particular, can be traced back to this assumption, and many authors have argued that these traditional criteria of scientific adequacy are not applicable to the human sciences (e.g. Ravetz, 1971). The dangers of regarding certain scientific traditions as dogma are very real.

Much of a scientific education merely serves to re-inforce these various attitudes. (Medawar, for example, specifically criticizes the public schools for fostering the idea of the superiority of pure over applied science (ibid., p.17).) Indeed, there are fundamental difficulties in teaching science in a way which achieves the correct balance between authority and the development of powers of judgment. If a pupil or a student is to emerge from his education as a potentially useful scientist, then it is essential that he absorbs a considerable proportion of the vast wealth of scientific knowledge which now exists. In the more traditional methods of education, this was achieved largely through the authoritative communication of neatly packaged knowledge. Theory was presented as a sequence of elegant proofs, and the trial and error fumbling which originally derived them, and which is a fundamental part of the scientific method, was largely ignored in school and undergraduate curricula. Practical experiment in the laboratory all too often became, not a search for knowledge, but an exercise in producing the 'right' result in order to obtain the marks necessary to overcome that stage in the academic obstacle course. Only when he began his own research career did the young scientist begin to acquire and appreciate the craft skills which are an equally important part of the scientific method.

In recent years, attempts have been made to reform syllabuses in order to lessen this degree of authority and instil a more rounded appreciation of the scientific method. Such reforms were clearly necessary if the student was to emerge as anything other than a walking textbook. They have not been wholly successful. Indeed, the worst have gone to the opposite extreme, producing highly imaginative scientific illiterates instead. But at least the dangers of both extremes are now clearly recognized, and there is hope that some of these fundamental problems can be solved in the future. Whether or not these new methods are successful, however, they will not alter the attitudes of today's scientists, many of whom went through an old-fashioned training, with its characteristic attitudes to authority and elegance.

Why are many scientists so unwilling to enter into political debate? The discussion so far provides one important clue. The prestige of a scientist rests, he believes, on his scientific knowledge and his ability to make accurate predictions based on that knowledge. Moreover, when he speaks, he speaks on behalf of science and not as an individual. If he permits himself any loose, inexact or ill-founded statements, then he will be attacked by the rest of the scientific community, accused of lack of rigour, and his reputation will suffer accordingly. Unfortunately, it is in the nature of social problems that, in order to suggest solutions to them, one must make a leap from the available facts to the possible solutions; such problems are so complex that, as yet at least, they are not susceptible to rigorous analysis. A scientist may be unwilling to make this leap because it oversteps the allowable criteria on which his professional reputation rests. Within science, he may put forward a hypothesis but, provided he clearly labelled it as tentative he would remain within bounds. Social problems, however, require practical solutions for immediate action, and the scientist has only two choices: either to say 'I don't know, science does not tell me', or to overstep the constraints of his profession and make value judgments, or judgments of fact for which some of the facts are lacking, for which, if he made equivalent judgments within his profession, he would be soundly condemned by his peers for ill-considered and precipitate action.

One might almost say, in attempting to summarize this argument, that science is an attempt to replace the inexact by the exact and, in so doing, remove value judgments as far as possible from consideration. Moreover, a scientific education reinforces this attempt to eliminate value judgments. Small wonder that scientists are unwilling to break the rules and enter into the market place.

Should a scientific education be modified in any way to overcome these drawbacks? The problem with the present system is not that it fails to produce good scientists, though Einstein, if he were alive, might have made some penetrating criticisms of this remark, but that the scientists it produces are unwilling to help to the best of their ability in the solution of social problems. The modern world, to quote Sir Hermann Bondi, 'requires above all adaptable human beings with intelligent minds, capable of tackling the vast number of different matters that come their way, as citizens of a changing nation, as people performing changing jobs in a changing world.' Not merely adaptable, but wise too. The point must be repeated. Some specialist concentration in the teaching of science is inescapable – yet it can have dire consequences. It is a familiar comment, but its familiarity should not conceal its importance.

One modification might be the more widespread inclusion of the history of science in a scientific education, which would reveal the way in which scientific experiments have historically always been conducted and interpreted within a wider framework of ideas which is itself in part derived from the authority of a 'consensus judgment'. Such a study might expose such hidden connections and generate an awareness of the relativity of our own judgments.

Of greater value, perhaps, might be the inclusion of a study of science within its social context. One of the fundamental drives behind the scientific revolution of the past two hundred years has been man's desire to manipulate and control his environment. Yet much of science, and in particular the pure sciences – chemistry, physics etc. – have been taught with little reference to the social and economic consequences of their practical application. It would seem ethically wrong to ignore these consequences and ethically desirable to teach something of them and to evoke an awareness of the moral considerations involved in, say, biological engineering or the uses of science in war. More than this, however, the teaching of science within its social context might well serve to produce that kind of human being which, as Sir Hermann Bondi points out, society needs in order to survive and adapt to changing circumstances.

Finally, there is a need to include ethics within the teaching of science. This does not mean that every scientist should have a course on formal ethical theory as a part of his training. Rather, it means that ethical issues which arise naturally in the course of a scientific education should be picked out and examined, and not merely brushed aside. Using live animals in teaching experiments provides an example. Such a use of animals needs very careful justification (if indeed it can be justified at all when it merely forms part of an undergraduate teaching programme and is likely to be repeated from class to class and from year to year), but such issues are rarely discussed with the students themselves. It is not uncommon for the fledgling research student to be shown, with clinical detachment, the correct way to kill a mouse, without any attempt on the part of the teacher to justify the killing of animals in such experiments. Similar remarks may apply to defence research. A teacher may have decided to undertake defence research only after long and careful agonizing over the issues involved. If he fails to discuss these with his research students, however, then he may, wittingly or unwittingly, give them the impression that there are no ethical problems in such work (Paskins, 1975).

Ethics is not a word which is often used in association with the teaching of science. Indeed, a scientific training contains within it an attempt to convince the student that the practice of science can be aloof from value judgments altogether. The unfortunate consequence of this approach is that scientists tend to give insufficient weight to ethical values. If the primary aim of education is the knowledge of values, and the production of wise and adaptable human beings, then the discussion and teaching of ethics must form one of its cornerstones, and not remain hidden away in the attic.

References

Glass, B., *Science and Ethical Values*, Oxford University Press (1966).

MacCallum, M., 'Gravity Experiments', *Nature*, 264 (1976), pp.14–15.

Medawar, P. B., *The Art of the Soluble*, Penguin Books (1969).

Paskins, B., 'The Scientist's Research and the Military in a Western Democracy', communicated to the Science and Ethics Committee (1975).

Ravetz, J. R., *Scientific Knowledge and its Social Problems*, Oxford University Press (1971).

11 Selecting Research

Problems worthy of attack
Prove their worth by hitting back.
Piet Hein

How does a young scientist choose a topic for his research? It might seem at first sight as if ethical considerations played little or no part in this decision. Surely, one might argue, the only really important criterion is the scientific importance of a particular project? The scientific community will judge his results by this criterion, together with the skill and rigour with which it has been approached, and it is on that judgment that the young scientist's future career will depend. Despite this, however, his choice will usually involve other values, and may have ethical implications.

There may be a strong element of chance. The laboratory in which he finds himself may specialize in one particular area of research rather than another, reflecting the interests of the scientists already working there, and will have accumulated not only a reservoir of technical expertise but the equipment necessary for such research. The current contraction in scientific research, and the funds available to it, is reducing the degree of mobility of the young scientist, so that his choice of research is becoming more dependent on the facilities which are available to him in whatever institution is prepared to offer him a post. Even within a particular area of research, a problem may crop up by accident, and not be consciously chosen. Such decisions have negative ethical implications. Allowing them to be determined by chance is, consciously or unconsciously, a rejection of ethical or other values as determining factors.

Alternatively, research may be seen as a means to a personal end. Most research scientists are seeking not merely to advance science but to advance their own professional standing as well, whether it be by publishing a sufficient volume of creditable results or simply acquiring the qualifications of MSc or PhD. In extreme cases, a supervisor may have to decide whether to direct a student to an important piece of work which may lead nowhere, or to a safe but comparatively trivial PhD problem. In all cases, he must recognize these two poles and find a compromise position which is suitable for that particular student. Here the decision, whether it is taken by the supervisor or the student or both, has ethical implications, particularly in the choice of 'end'. The student may make his 'end' the PhD, and eliminate other considerations. This, however, is a limited 'end' and may itself be determined by further 'ends' – to obtain a university post, or a research post in industry – which are themselves determined by other 'ends' – to make money, or find security, or teach the young, or add to the sum of knowledge, or be of practical service to mankind, or win a Nobel Prize, or have intellectually stimulating work, and so on. Moreover, the means chosen to reach the first goal of the PhD may affect the possibility of reaching the second goal. Choice of an unadventurous research area may produce a dull teacher, or someone who is wedded to a safe but unexciting branch of science. It may even produce disenchantment with research as a whole and lead to the abandonment of the original long-term objective of a career in research. Alternatively, choice of an exciting but uncertain area of research may stimulate someone into replacing his original goal with the desire for a career in research, even if this necessitates a lessening of his previous aspirations for, say, a high salary.

Similar remarks apply throughout a scientist's career. He may 'go for the big one' and choose a particularly difficult problem, knowing that success will bring considerable personal prestige, whilst failure may leave him socially worse off than if he had chosen a more tractable, but less prestigious, area of research where he is reasonably assured of some results. The pressure to produce results, and the social stigma attached to failure, are just as prevalent in academic research as in any other highly competitive profession.

To an outsider, the fact that much of the impetus for the advance of science arises from the status-seeking of individual scientists might seem unfortunate. Indeed, in view of the protestations of such scientists that the essence of science is a disinterested search for the truth, it might almost seem to border on hypocrisy. Perhaps it does. But one of the reasons why science has been so successful over the past two centuries is because it has developed social institutions which 'cunningly sublimate human ambition and competitiveness into the search for new knowledge' (Wren-Lewis, 1974). Where it becomes unfortunate is not that it allows the individual to sublimate his personal ambitions, but that it may lead him to believe that he can submerge his whole personality, warts, human failings and all, in science, leaving only a dehumanized craftsman above the surface. Here again, the practice of science may reinforce the scientist's mistaken belief that science has nothing to do with ethics.

Another factor which influences a scientist's choice of research is his interest in it. Intellectual stimulation, and the personal satisfaction which research produces, are 'ends' of research which are just as important as any others, and their achievement depends upon the personal interests of the individual. It is considerations of this type which separate theoretician from experimentalist, or botanist from zoologist. Clearly, choice of a research area for reasons of personal interest again involves a value judgment, since it allows personal interest to have a higher value in choosing a topic than the intrinsic scientific or social importance of the work. Moreover, they are judgments which mirror many of the values which are fundamental to science: respect for particular research because of its 'beauty' or rigour, quite independently of its value to the rest of science or to society as a whole.

However, there are cases where what are normally regarded as ethical considerations explicitly enter into the selection of research. A scientist may, for example, deliberately go into, or refuse to go into, military research, because he is a patriot or believes in the power of deterrents, or because he is a pacifist or believes that there is a disproportionately high expenditure on defence. He may go into, or refuse to go into, industrial research, because he believes that industrial research will improve the standard of living of ordinary people, or because he is opposed to the capitalist system. These are ethical decisions. He may choose to go in for 'pure research' as a conscious or unconscious way of escaping from this sort of consideration; that too is a decision with ethical implications. He may accept a post in military or industrial research simply because it is offered, without worrying about the nature of the work involved or the system behind it; to do so is to give implicit ethical approval to that work and that system. In every case, whether he recognizes it or not, the working scientist is taking an ethical decision.

There is one notable omission from the list of reasons given so far. Does a scientist's choice of research topic depend, in any way, on the social benefit which he anticipates will result from that research? In some cases the answer is obviously yes. Much of industrial research is a conscious attempt to produce a product with certain specifications, whether it is a new drug, washing powder or automobile. But scientists within industry are more clearly the servants of social goals than their academic counterparts; it is in the nature of their employment that they are not wholly free to choose their area of research but are constrained by the needs of their employer and of society as a whole.

These constraints are largely absent from university research. Indeed, many academic scientists wholly ignore considerations of social utility in choosing their research programme. Science, they would argue, is the free and unfettered search for knowledge, and not merely a tool for social improvement. Moreover, scientific knowledge is an indivisible whole, so that concentrating solely on those areas of research which one might expect to produce social benefits may well, in the long run, delay their arrival. It is impossible to predict, they would argue, where solutions to particular problems may be found; they may turn up in the most unexpected places. There is some truth in this argument, but it can be carried to extremes. Thus, in one article, Isaac Asimov attempted to justify sending men to the moon by arguing that it might, just might, lead to a cure for cancer (quoted in Dixon, 1976, p.135). Obviously, this kind of argument can be used to justify any research, however esoteric.

Yet, despite its flimsiness, scientists do use this kind of argument to justify their rejection of criteria of social benefit in selecting research. Indeed, at times, this rejection becomes a hallowed principle. This is what the 1965 AAAS report, already mentioned, has to say on the subject:

> Too often science is regarded only as a means of satisfying immediate social demands, and such demands sometimes produce pressures which erode the integrity of science. Society must recognise, more clearly than it now does, that such pressures are self defeating, and, given the hazards involved in a faulty understanding of the powers of modern science, exceedingly dangerous as well (p.197).

Obviously, one can never be certain that a particular piece of research will produce the anticipated social benefits, nor that the key breakthrough will not be made in some totally unrelated, and apparently socially useless, branch of science. But then to assume that, because of this, criteria of social benefit become irrelevant to the planning of research programmes is totally unjustified.

Debates of this kind, between the desire of the scientist to pursue the research of his own choosing, and the desire of society for social benefits from science, are familiar enough. But, whilst it may be true that society places too much emphasis on research where there is clearly a possibility of immediate social benefit, it is also true that scientists give too little weight to the social relevance of their research. Both have unfortunate consequences. Thus the biologist may go to elaborate lengths to justify his proposed research in terms of the possibility that it may produce a cancer cure, quite irrespective of whether that is the object of his research or not, simply because funding is available for cancer research on a scale without parallel in the rest of biology. Grantsmanship of this type, which a scientist must practise if he is to be successful in obtaining large-scale financial support, only leads in the long run to cynicism and disillusionment on both sides. But it is equally true that much of the research carried out in universities at present is almost totally irrelevant to the needs of the nation either now or in the foreseeable future. If scientists continue to ignore this fact, they cannot complain if people begin to show signs of disillusionment with academic research as a whole and restrict funding accordingly.

Behind these arguments lie some very deep ethical problems concerning the purpose of scientific research, and the freedom which should be accorded to scientists. On the one hand, science may be regarded as equivalent to any other intellectual pursuit, and allowed the same freedoms. There is no suggestion in this country, for example, that poets or composers should be constrained by considerations of social utility, though this is in stark contrast with the attitude taken in other countries, such as Soviet Russia. Surely, scientists would argue, they should be allowed the same freedoms as their artistic counterparts? These arguments, however, are complicated by those special considerations which set science apart from other intellectual pursuits. Science is clearly of value to man; its applications have totally transformed our civilization within the last two hundred years. Moreover, our dependence on science has inevitably meant that scientific research has become a major political issue. As has already been emphasized, no country would be prepared to allocate so large a proportion of its total resources to a single endeavour, such as space exploration, for purely intellectual motives.

This is the nub of the problem. Academic scientists, like any other intellectual specialists, resist any encroachment on their intellectual freedom because, as they quite rightly argue, any such restrictions lessen their abilities as scientists. Yet governments will only be prepared to continue the present vast level of finance allotted to scientific research if they can be convinced that social benefits will arise from such research. The problem, as with so many ethical issues, is one of balance. Obviously, there is a place for pure science, unfettered by considerations of immediate social utility, just as there is a place for any other academic discipline. But, equally obviously, scientific research will only continue to receive the present level of funding, which is allocated to it for social and political motives, if scientists are prepared to accept the constraints implied by those motives. To demand that all scientific research should be manifestly socially useful would be an intolerable restriction on intellectual freedom. To demand that all science should be free of social pressures and constraints is to accord a position of privilege to science which a society, hemmed in by immediate economic and political pressures, can ill afford to do.

Whether the present balance is right or not is a matter of personal opinion. Many scientists undoubtedly feel that they receive too little financial support; many laymen feel that they get too much. But, leaving aside the major projects, such as space research, where social considerations must dominate the debate, one can argue that the present balance is fair. As Sir Alan Cottrell, formerly Chief Scientific Adviser to the British Government, has pointed out, the strains at present being felt in the country's scientific research organizations are not due to an increasing pressure for all science to be socially relevant but to the Government's efforts to hold down public expenditure as a whole. In his view, 'so far as basic science is concerned, the record of governments for steady, consistent, support, throughout many years of national economic difficulty, is good' (Cottrell, 1976, p.82).

There is another aspect to this debate. Financial considerations are not the only ones which 'interfere' with a scientist's freedom to choose research. Other considerations may also play a part. In vivisection, for example, discussed in chapter 3, freedom in research is explicitly constrained by ethical factors: the anticipated benefits from the research and the duty not to inflict unnecessary pain and suffering on animals. Considerations of human safety are even more important. One area of modern biology, the so-called recombinant DNA research or genetic engineering, has been the subject of much recent concern. A number of eminent scientists have proposed that there should be a moratorium in this field. Government committees, and even City Councils, have debated the pros and cons, and the correspondence columns of scientific journals have been full of it (see, for example, Culliton, 1976/2, Dyson, 1976 and references cited therein).

A moratorium in this field may or may not be desirable, but that is too complex an issue to be discussed here. It is relevant, however, that, even here, scientists try to wriggle out of ethics. Paul Berg, one of the main scientific advocates of the moratorium, explained at one conference that his decision had nothing to do with ethics. 'It is simply a public health problem', he said. This, as one leader writer in *New Scientist* pointed out, is an 'astounding position' (Lewin, 1974). One is reminded of the observation by the 1965 AAAS committee, quoted in chapter 5, that the assessment of degrees of risk is 'not a scientific conclusion, but a social judgement'. Any debate for or against a moratorium must weigh the potential benefits against potential risks which is, almost by definition, an ethical decision. Moreover, it is a decision which must be taken by society as a whole, or their elected representatives, and not just by scientists. And when there is disagreement amongst scientists, it can be very difficult for the layman to make a valid judgment. As Councillor D. E. Clem of Cambridge, Massachusetts put it: 'I tried to understand the science, but I decided I couldn't make a legitimate assessment of the risk. When I realized I couldn't decide to vote for or against a moratorium on scientific grounds, I shifted to the political' (quoted in Culliton, 1976/2, p.301).

This point can be emphasized (and, alas, in view of the attitudes of many scientists, it needs emphasizing) by quoting Senator Javits:

> The decisions with respect to the future of biomedical research, the determination of priorities, the weighing of the nonquantifiable social costs and benefits of medical technology – these decisions are in fact political because they involve the entire body politic including, of course, the research community itself. A scientist is no more trained to decide finally the moral and political implications of his or her work than the public – and its elected representatives – is trained to decide finally on scientific methodologies (quoted in Culliton, 1976/1, p.452).

To which one might add that, not only are scientists unskilled in making ethical and political decisions, but that many scientists are unwilling even to recognize that such judgments are ethical and are not their exclusive concern.

The scientist's desire for freedom in research is understandable and, to some extent, justifiable. He is the expert, and unnecessary interference in his craft by the layman will do nobody any good. But he cannot demand isolation from society. Decisions on public expenditure and public safety are public decisions. Science is a social activity, and the scientist is accountable to society as a whole for his behaviour. The writer can legitimately expect some public financial support and relative freedom to practise his craft. But he cannot demand unlimited resources, nor can he demand freedom from such social constraints as the laws of libel and obscenity. He is a part of the community, not aloof from it. The same is true of the scientist, and failure to recognize that fact may, in the long term, be detrimental both to the practice of science and the benefit of mankind as a whole.

References

AAAS, 'The Integrity of Science', A Report by the AAAS Committee on Science in the Promotion of Human Welfare, *American Scientist*, 53 (1965), pp.174–198.

Cottrell, A., 'The Rise and Fall of Science Policy', *New Scientist*, 72 (1976), pp.80–82).

Culliton, B. J., 'Public Participation in Science: Still in Need of Definition', *Science*, 192 (1976/1), pp.451–453.

Culliton, B. J., 'Recombinant DNA: Cambridge City Council Votes Moratorium', *Science*, 193 (1976/2), pp.300–301.

Dixon, B., *What is Science For?*, Penguin Books, (1976).

Dyson, F. J., 'Costs and Benefits of Recombinant DNA Research', *Science*, 193 (1976), pp.6–8.

Lewin, R., 'Ethics and Genetic Engineering', *New Scientist*, 64 (1974), p.163.

Wren-Lewis, J., 'Science Comes of Age. A Review of *The Sociology of Science* by R. K. Merton', *New Scientist*, 62 (1974), p.563.

12 Solving Social Problems

No problem is too difficult to be solved by a theoretician.
Persian Proverb

In the last chapter, an attempt was made to show that social and ethical considerations cannot be excluded from the business of science. The discussion was restricted to one topic, namely the choice of research project, but this type of analysis can readily be extended to other aspects of the practice of science: in chapter 2, for example, the ethics of publication were discussed in some depth. The purpose of this chapter, and the next, is somewhat different. For, whilst some problems arise because ethics impinges directly on science, others arise whenever science is applied to social problems. To understand this second group of problems, it is necessary to explore the extent to which the scientific method, as it is currently understood, is capable of solving social problems, and where it breaks down and only adds to the confusion. If the limitations inherent in the application of scientific techniques to social problems are not fully recognized, both by the scientists themselves and by society as a whole, then there is a real risk of falling into one of two traps: either assuming that science can solve all problems, or denying that science has any contribution to make. At present, there seems to be more danger of falling into the former rather than the latter, but either could be detrimental to the well-being of society and the progress of science.

The scientific method is a curious collection of techniques, involving as it does both a mixture of guesswork, rhetoric and intuition, and a reliance on careful mathematical rigour, reproducibility and attempted objectivity. As a subject for philosophical enquiry, it has received a number of penetrating analyses in recent years (see, for example, Toulmin, 1960; Kuhn, 1962; Ziman, 1968; Ravetz, 1971 or the work of Karl Popper). Working scientists, however, whilst they might read such accounts with enjoyment and interest, are rarely guided by them. Their grasp of the scientific method is intuitive, based on a scientific training and a day by day contact with the corporate body of science and the standards which it has evolved, and not on a reasoned analysis of the underlying principles. In this they are not unlike their artistic counterparts, many of whom show a healthy scepticism, occasionally bordering on contempt, for the analysts – musicologists, literary critics and the like – who seek to establish the patterns behind artistic endeavour rather than to create fresh works of art themselves.

Of course, the scientific craftsman does not need to understand the theory behind his craftsmanship, any more than one needs a thorough grasp of theology in order to be a good Christian. But it can lead to problems when intuitive beliefs are applied to situations where they may not be wholly applicable. The methods of science have been developed and refined over the past few centuries to deal with those problems which arise from the study of the natural sciences – principally the physical sciences, but, more recently, the biological sciences as well. These problems, and the techniques which are used to solve them, show many special features, some of which are discussed below, which may not be applicable to other types of problem.

In his thorough discussion of this subject, Ravetz has identified three types of problem: scientific problems, such as why an apple falls to the ground; technical problems, such as how to build an aircraft; and social (or what Ravetz calls practical) problems, such as how to decrease unemployment or urban violence (Ravetz, 1971). Each type of problem is different, involving its own set of techniques in the search for a solution, and its own set of standards in assessing the adequacy and success of any solution which is proposed. And, of more importance to the present discussion, the techniques and criteria which are suitable for one class of problems may be wholly inappropriate to another.

No chapter of this length can attempt to explore the similarities and differences between these three types of problem in any depth. But even a cursory glance at the subtleties involved will help to highlight the difficulties inherent in trying to apply the scientific method to social problems, and to suggest why scientists have such a poor record in solving such problems.

Social problems, according to Ravetz, are tackled in five stages: definition of the problem, analysis of the many facets involved, drawing conclusions from that analysis and framing decisions around those conclusions, execution of those decisions, and monitoring and controlling the results so that the desired goal can be achieved (ibid., p.340). This kind of description tends to under-emphasize the important role played by social principles in the approach to social problems but, nevertheless, it provides a reasonable model for the process used in practice. Two aspects of this sequence need discussion here. The first is the process of analysis, which is the traditional role filled by the scientist. The second is the extent to which the citizen scientist should play a role in the later stages of this exercise, when decisions have to be made and executed.

Consider a typical scientific problem. Why does the apple fall? The aim of the scientist, in trying to answer this question, is to find an explanation which covers a wide variety of other observations, one which is compatible with his explanations of other types of event, and one which he can use, with some assurance, to predict future events. To do this, he will make a number of other observations, to see, for example, whether other objects fall if released in the air. He may find that he has to set certain facts on one side (such as flying birds) in the hope that he can cope with them at a later stage. He may isolate his problem, and assume that it is independent of other factors. He may check, for example that the falling of apples is independent of the time of day, but he may not check which way they fall in Australia. He may introduce new concepts such as weight, force or gravity. Ultimately, he will reach, or hopes to reach, a generalized statement about the relationship between these concepts, expressed in the language of mathematics, which explains his original observations in the sense that it is compatible with them all (or is sufficiently all-embracing that he can justifiably hope that the exceptions can be fitted in later when more knowledge is discovered), is aesthetically and philosophically pleasing, and can be used to predict future events reliably.

This is, it must be admitted, a gross oversimplification of the scientific method, but it does bring out some of the essential features. Firstly, it is wholly concerned with the world of physical objects. Secondly, it proceeds from a wealth of observable data through processes of measurement, simplification, generalization and isolation of problems, and that each stage involves certain assumptions which may or may not be explicitly stated, and whose validity can only be checked by the success of the final result. Thirdly, it devises concepts which, whilst they are strictly related to observable data, are nevertheless intellectual abstractions which are not a part of everyday experience. Fourthly, the goal of the scientific endeavour is the construction of mathematical relationships between those concepts, whose adequacy is judged by their compatibility with each other and with the data from which they are constructed, by their elegance and aesthetic appeal, and by their predictive power.

The success of these methods as applied to physical and biological problems is too familiar to need elaboration here: this has been the most significant achievement of the nineteenth and twentieth centuries. Moreover, it is because of this success that many people have assumed that these methods could be applied with equal effectiveness to problems of human behaviour. But the human sciences – economics, sociology, psychology, and political science – have not yet begun to approach the successes achieved by the physical and biological sciences. Why this is so is a very deep problem which, as yet, cannot be answered. Perhaps, as some of the scientists working in these disciplines believe, it is simply a question of time: the application of scientific methods to human problems is still relatively recent, and its failures may merely arise from inexperience which further efforts will overcome. But it is also possible that these traditional methods are inappropriate to human problems, or that the assumptions on which they are based simply break down in the human world. Certain assumptions on the uniformity, regularity and repetitiveness of the natural world are fundamental to the application of the scientific method to physical problems. One assumes that, for example, if one apple falls, then, under similar conditions, carefully specified, all other apples will fall in a similar way. There is no guarantee that similar assumptions can be applied to the human world; if they do not apply, then the application of the scientific method becomes much more difficult, if not impossible in principle.

Nevertheless, scientists have approached social problems in the belief that traditional scientific methods can be used. Since this fact has not yet been established, then the results of such sciences must be treated with considerable caution and the disciplines themselves can, following Ravetz, be classified as immature (ibid.). What is more important, however, is that society looks to these disciplines to clarify its social problems. Under these circumstances one can talk of a folk-science, since the role of that discipline is very similar to the role of other folk-sciences in the past (ibid.). People use it to justify certain courses of action, or merely to analyse problems and suggest solutions, even though the results of the discipline are still disputed by its practitioners and will not stand up to the rigorous scrutiny which would be normal in the natural sciences.

Whilst these judgments may seem rather harsh, they are not intended to be pejorative. Real solutions to social problems can only be found if the same depth of understanding can be achieved as has been achieved for the physical world and, in the absence of any other criteria, we must move forward by trial and error. But there are two very important dangers to be avoided. One danger is to forget that this approach is based on fundamental, untested assumptions, and to assume that preliminary results have the same truth-status as preliminary results in other disciplines. Failure to keep these assumptions in mind can so easily lead to the construction of elaborate 'castles in the air': theories which have no real correspondence with reality. The second danger, which follows on from the first, is to base social policies on these results as if they had the status of scientific 'truths'. They may have no more validity than the leeches of folk-medicine, or they may be as effective as some herbal remedies. As yet, we simply do not know.

The analysis stage of social problem-solving is, for these reasons, riddled with difficulties and dangers. Each of the traditional scientific tools will be used to produce self-consistent results, but the real validity of those results must remain in doubt. Detailed consideration of these difficulties will be deferred to the next chapter, which will consider one such tool in some detail. But a brief look at other examples may help to highlight these problems.

One way in which a scientist tries to come to grips with a particularly elusive concept – the hardness of rocks or the reactivity of chemicals – is to devise some measurable parameter which, in some way, parallels the quality under consideration. If he can do this, and this parameter can be fitted into a broader theoretical framework, then his understanding of the original concept is thereby enriched, if not improved. At its most successful, the original concept may be wholly replaced by a new parameter or parameters. Inertia, for example, is not a concept one commonly meets in textbooks on mechanics; instead, they talk about mass, force and energy.

The danger which arises when this technique is applied to social problems is that the parameters devised seem to take on a life of their own. The 'quality of life' is a very elusive concept which one attempts to pin down by devising parameters, such as 'gross national product' or 'income per capita', which are, in some way, related to it. Then, even though the parameter may be ill-defined (even such relatively simple concepts as 'price' and 'profit' are notoriously difficult to define rigorously (see Ravetz, 1971, p.112)) and attempts to fit it into a broader theoretical framework unsuccessful, nevertheless the aim of social policy becomes the maximization of the parameter. The successes and failures of this approach are well known.

Similar difficulties arise when scientists attempt to isolate problems. One essential stage of the scientific method is the search for those parts of a problem which can be isolated from their surroundings, and those parts which cannot be so separated but form an indivisible whole. Any isolation of a part is an approximation, but one which need not affect the result. Sending rockets into space, for example, undoubtedly affects the time it takes the earth to go round the sun, but the effect is negligible. On the other hand, attempts to formulate a theory of earthquakes which ignores the motions of the other planets may, if recent suppositions prove correct, inevitably fail (Gribbin and Plagemann, 1974 though see also Hughes 1977).

Attempts to isolate social problems, by assuming that certain quantities would not affect, or be affected by, the particular problem under consideration, are fraught with difficulties. Unfortunately, one can only suggest solutions to many real problems by isolating them in this way and, since the isolation is artificial, the solution which seems obvious if the problem is considered in isolation may have undesirable consequences when applied to the real problem and, indeed, may prove to be no solution at all.

It is not hard to find examples of this. Isolating the problem of famine from other social problems makes the solution seem obvious: more food. In practice, however, as has already been discussed, increased food production has not solved the problems of world famine but has had other, possibly undesirable, social effects. Similar problems arise in economics. Are the solutions to our present economic difficulties really as simple as some politicians would lead us to believe (print less money, cut down public expenditure and so on)? One has one's doubts.

The production of undesirable side-effects from particular courses of action is even more common. Detergents not only mop up oil slicks but kill fish as well. DDT not only kills the malaria-carrying mosquito but, if it is allowed to accumulate in sufficient quantities, most other animal species as well. The newspapers are full of such examples. Indeed, whether one is considering the eco-sphere or the world of economics and politics, the main facts which stand out are the complexity of the total system, and the close interdependence of its constituent parts. It is not surprising that any modification to one part affects many others; it would be astonishing if it did not.

At the risk of falling into the same trap as many scientists seem to fall into, one might suggest that the main failing of scientists is to underestimate the complexity of social problems and to choose oversimplified approximations. Indeed, the search for simplicity is fundamental to the scientific method, and one of the pinnacles of scientific achievement is the ability to explain a wide variety of phenomena with as few laws and equations as possible. Einstein's formula, $E=mc^2$ is not just another scientific law; it has become a symbol of the whole scientific method. The belief that the complexities of the physical world can be reduced to a handful of basic laws is fundamental to the scientific approach, and, so far, this belief has been justified. Attempts to reduce the world of economics and politics to similar basic laws have, so far, been a dismal failure.

The difficult decision which faces society, and it is society as a whole which must make this choice and not the scientific community by itself, is what status to give scientific predictions, and what duties to demand of its scientists. Social forecasting is not unlike long-range weather forecasting: exceedingly difficult and notoriously unreliable. Unfortunately, whilst long-range weather forecasts are sensibly couched in the vaguest language, social forecasts are often made in precise, numerical language which, as Lord Ashby has pointed out, gives them an unwarranted air of accuracy, not to say authority (Ashby, 1975, p.4). *The Limits to Growth*, a typically gloomy set of scientific predictions which has acquired considerable notoriety in recent years, falls into just this trap. Despite the shaky nature of many of its assumptions, its conclusions, based on computer simulations, are presented with careful and misleading accuracy.

Is it ethically wrong for scientists to make this type of prediction? Scientists undoubtedly have a duty to use their expertise to explore social problems and their possible solutions. They also have a duty to emphasize that their hypotheses and opinions should not be regarded as authoritative predictions and should, for that reason, avoid precision in prediction unless it is accompanied by equal precision in error and uncertainty estimation. But, and this is the important point, what happens when society bases its policy on scientific opinions which subsequently prove to have been mistaken? Is society to blame for attaching too much importance to the pronouncements of scientists? Are scientists to blame for giving opinions when they have good reason to suspect that society will attach too much weight to their opinions? Does it have any meaning, or any purpose, to attach blame to either side?

Put it another way. If society attributes an authority to such predictions by scientists, does that necessarily involve the scientist in special responsibilities, even if he disclaims that this predictions should carry any more weight than those by ordinary people? Those who seek authority must also assume the concomitant responsibility, but what of those who have authority thrust upon them?

References

Ashby, Lord, 'A Second Look at Doom', Twenty-first Fawley Foundation Lecture, University of Southampton (1975).

Gribbin, J. and Plagemann, S., *The Jupiter Effect*, Macmillan (1974).

Hughes, D. W., 'Planetary alignments don't cause Earthquakes', *Nature*, 265 (1977), p.13.

Kuhn, T. S., *The Structure of Scientific Revolutions*, University of Chicago Press (1962).

Meadows, D. H., Meadows, D. L., Randers, J. and Behrens, W. W. III, *The Limits to Growth*, Pan Books (1974).

Popper, K., *Conjectures and Refutations*, Routledge (1963).

Ravetz, J. R., *Scientific Knowledge and its Social Problems*, Oxford University Press (1971).

Toulmin, S., *The Philosophy of Science*, Harper & Row (1960).

Ziman, J. M., *Public Knowledge*, Cambridge University Press (1968).

13 Measuring the Alternatives

There are two problems in my life. The political ones are
insoluble and the economic ones are incomprehensible.
Sir Alec Douglas Home

One of the reasons why scientific statements seem so authoritative is
the precision with which they are expressed. The scientist does not
merely say 'if water is heated, it will eventually boil', or 'people need
Vitamin C', he is able to state the exact figures in each case, and the
conditions under which they apply. Indeed, the development of
probability theory and the theory of errors allows scientists to
quantify even their own uncertainties. As one author put it 'even
when a scientist acknowledges a lack of precision, he seems
paradoxically confident in defining his own imprecision' (Friedlander, 1972, p.23).

There are difficulties, of course. Some concepts, such as the
electronegativity of a chemical element, seem to defy accurate
definition and measurement. In other cases, it may be necessary to
develop a new concept which is accurately measurable but does not
correspond exactly to the way in which we experience the world; the
concept of energy is a good example. Then again, there may be
fierce debate on whether the numbers one obtains actually measure
the concept one is trying to pin down. Do IQ values really measure
intelligence? Many scientists think they do not.

Despite these difficulties, however, the expression of scientific
statements in the precise language of mathematics gives them an air
of confidence and authority which is lacking in ordinary statements,
an aura which is reinforced by their predictive power. Not unnaturally, many scientists, and others, wish to extend the scientific
method, and the certainty and authority which it brings with it, into
the world of value judgments in the hope that many of the problems
which are associated with value judgments can be removed. Some
of the general limitations and dangers of this approach were discussed in the previous chapter; the present chapter will highlight
these issues by considering one particular technique.

Making social decisions is a complicated business. Democracy,
which, according to Bernard Shaw, 'substitutes election by the
incompetent many for appointment by the corrupt few', is in one

sense a mechanism by which social action is governed by what the majority of people want. But one might equally argue that 'a political election is as much a ritualistic exercise for the integration of society as a procedure for deciding who shall govern' (Ebling, 1969, *Introduction*, p.xxvii). If the majority are prepared to support a particular policy, and the opposing minority are prepared to accept the majority verdict, then reaching that policy by democratic means is much less important. Indeed, any attempt to take complex decisions by an electoral process falls foul of 'the problem of deriving a collective preference from a set of individual preferences' (Phelps, 1973, p.9). In extreme cases, it may be expedient to hold a referendum, as in the case of the Common Market. But every decision cannot be taken by referendum; if it did then effective government would cease. What is more, referenda can be used only if the issue can be boiled down to a simple question; in more complex situations a unique collective preference may be impossible to define. If there are three choices, for example, then the voting paradox of Condorcet points out that, under certain conditions, there may be a majority which prefers A to B, a similar majority which prefers B to C, but an identical majority which prefers C to A (ibid., p.18). This becomes even worse in more complex cases. We may decide that there should be a third airport for London but, since most people want it to be built as far away from them as possible, the result of a referendum might be that there is nowhere to build it (except in the middle of the North Sea!).

The essence of the scientific method is to replace qualitative judgments by quantitative ones, thus minimizing uncertainty and guesswork in making decisions. For this reason, considerable effort has been expended in recent years to try to find a sensible way to measure the various alternatives involved in social decisions. Since money is the mechanism by which most countries relate the values which they attach to commodities and services, it seems natural to try to express other values in monetary terms as well. This type of analysis does not, of course, remove the need to make social decisions. 'Cheapest' does not necessarily equal 'best', however thorough the analysis, nor does it necessarily equate with what people actually want. It can clarify problems and, like the appeal to majority vote, can provide principles and mechanisms with which it is possible to offer some justification for particular decisions. But there are dangers. Numerical analysis can be used to give spurious support to particular policies: support because the very precision of mathematical arguments gives them persuasive power; spurious, because the assumptions underlying this analysis may be inapplicable. Indeed, this kind of analysis can do more harm than good, actively distorting the issues involved and not clarifying them. This is one of the reasons why some people, including one member of the present Science and Ethics Committee, argue that attempts to quantify all values are undesirable and should be avoided.

Cost-effectiveness techniques typify this kind of analysis at its simplest level. Suppose that society has agreed on a particular goal, but that there are various alternative routes by which it could be reached. One way in which these various alternatives can be compared is to express the variables involved in monetary terms to determine which alternative would be the cheapest.

The health service provides many examples where this type of technique has been employed. Chronic renal disease, which ultimately results in irreversible kidney failure and death, cannot be prevented at present, it can only be treated, either by renal dialysis or by kidney transplants. Unfortunately, such treatment is very expensive. 'It has been estimated, for example, that to extend the renal dialysis and transplantation technique to all those who would benefit by its application in France would cost as much as the whole of the rest of the French Health Service' (Miller, 1975). And yet, without such treatment, many people would die. Indeed, one estimate for the United States put the number of people who die each year from kidney failure, whose life spans could be appreciably prolonged by the appropriate treatment, as high as 6000 (Klarman, Francis and Rosenthal, 1968, p.230). Obviously, the relative long-term costs of dialysis, as opposed to transplants, is an important factor in deciding which treatment to use. Cost-effectiveness studies have indicated that transplants involve less cost per year of increased life-expectancy than dialysis, suggesting that transplantation should be the preferred technique (ibid., p.239).

However such calculations, based on quantities like the costs of doctors' or nurses' time, or the cost of drugs and machinery, are notable more for what they leave out than what they include; simple financial considerations are not the only ones which are important. For example, it does not take into account which technique gives the longest life-expectancy for each individual (though as it happens, transplants win by this criterion as well). Nor does it consider the ethical problems of providing a supply of kidneys for transplants. Should there be, for example, a contracting-out principle, or should every individual consciously have to decide to leave his kidneys to medicine in the event of his death? Is it permissible to use live donors? One author at least has no doubts:

> What is the surgeon proposing to do to the kidney donor, who is
> his patient and stands in a professional relationship to him? He
> proposes to put his life at risk in a major operation in which he
> intends to maim him. There is no idea whatever of providing his
> patient with any medical benefit, but exactly the opposite, to do
> irreparable damage which may at any time become critical to his
> life. . . . (The) claim that no donor is accepted as such if it is shown
> that he has been under any emotional pressure, will carry no
> conviction to a psychiatrist . . . where other relatives have not
> sought to influence a man, he is quite capable of putting an unfair
> pressure on himself. It is all too easy to 'blackmail' oneself into
> unwise action out of a mistaken sense of where one's duty lies. . . .
> The economics of transplant surgery should also be looked at
> from the ethical point of view. What we are doing is to cannibalize
> the new model to keep an old crock on the road. Quite clearly,
> this is not the way to tackle the problem of obsolescence
>
> (E. T. O. Slater, in Ebling, 1969, p.106).

But this analysis, however valid, completely ignores the courage of
the donor who risks his life for his friend. To dismiss courage and
self-sacrifice as 'unwise action', arising out of a 'mistaken sense of
duty', is, some people would argue, to debase the human spirit.

The balance sheet for kidney transplants, therefore, still seems to
contain two types of quantities. On the one hand are those quantities which can readily be expressed in financial terms – salaries and
costs of equipment – whilst, on the other, are ethical values. Obviously the recipient of a transplant benefits from it, whilst the donor
suffers physically, though his feeling of mental well-being in 'doing
a good deed' is clearly of benefit to him and cannot be left out of the
calculation. How can one attempt to discover whether there is a net
benefit to society, or a net loss, from transplants? Does the recipient's gain outweigh the donor's loss? One obvious way to try to
answer such questions is to express *all* items on the balance sheet in
the same currency.

This is what cost-benefit analysis attempts to do. Unlike cost-
effectiveness techniques, it attempts to put a monetary value on all
parameters in the problem, including the benefits that arise from a
particular course of action, as well as both its financial and human
costs. Obviously, this bristles with problems. If the benefits are
simply financial savings, then this is relatively easy, though a glance
at the literature will show that, even then, the economist has many

difficult technical problems to solve (e.g. Layard, 1972). When the benefits involve savings in time, or reduced risks to life and limb, or the anticipated spin-off from research programmes, then these difficulties become acute. Benefits can only be compared objectively by translating them into common monetary units, but how to do this in many cases is far from clear.

To take a particularly difficult problem, can one put a price on human life? Economists have not been shy in coming forward with answers, and there is a considerable literature on the monetary value of human life (e.g. articles in Cooper and Calyer, 1973, or in Layard, 1972). However, one's first reaction on dipping into this literature is one of disbelief. Can it really be true that 'the most common way of calculating the economic worth of a person's life . . . is that of discounting to the present (i.e. estimating at current prices) the person's expected future earnings' (Mishan, 1971, p.219)? To quote an old rhyme: 'If life were a thing that money could buy, the rich would live and the poor would die.' This seems to be the ethical principle behind such economic arguments. Nor do the more sophisticated methods of valuing human life entirely avoid this difficulty. It is well known that cost-benefit analysis at present has an inbuilt bias in favour of the relatively rich and against the relatively poor. The values which any cost-benefit study use depend upon the existing distribution of income and wealth, and thereby contain an implicit value judgment that this distribution is ethically fair. Cost-benefit analysis cannot estimate our relative values for us.

What it can try to do, however, is attempt to discover how society compares different values and balances them against each other. The argument that the economist has no right to set a value on human life is, many economists would argue, misconceived. As Wolfe put it, 'it is not the economist, but society, which evaluates such matters; the economist can do little more than impose consistency on such social evaluations' (Wolfe, 1973, p.11). Many critics reject cost-benefit analysis because the monetary values it has produced so far for items like the value of human life are unacceptable. Henderson, for example, in a discussion on cost-benefit considerations in traffic safety, concludes that 'human life can be assigned a quantitative value for this purpose, but because of severe problems in measurement, involving over-riding value judgements which can affect the pricing by several orders of magnitude, it is a necessarily arbitrary and usually controversial technique' (Henderson, 1975, p.19). Such criticism is misplaced. Cost-benefit analysis does not produce misleading values because it is a 'necessarily arbitrary' technique, but because the problems it is trying to solve are very difficult, and economists have yet to refine what is still a very young technique to the point at which the values it produces afford real clarification of social problems. The history of science is full of similar cases. Early attempts to use quantum mechanics to predict the energies of chemical reactions gave answers which were wrong by several orders of magnitude, but no one rejected the technique as 'necessarily arbitrary'. To judge a theory by the skill of its practitioners is really rather silly.

Using a person's future estimated earnings gives a wrong result for the value of his life simply because the economist, in trying to use cost-benefit analysis to solve the problem, has made a number of false assumptions and misleading approximations. His theory is sound, but he lacks the skill to produce exact answers, and he has yet to find the right approximation which allows him to produce answers which are not too far from the truth. And it is not hard to see why it is so difficult to discover what value society places on human life. You and I may have insured our lives for £10,000, but this does not represent its value to us. And what of the millionaire who pays a high ransom for his kidnapped son, or the Apollo-Soyuz linkup, ostensibly an astronautical lifeboat, which must have cost over $500,000,000. Is this the value we place on the lives of astronauts? Even if it is, how can one 'impose consistency on such social evaluations'? Most people, when faced with the prospect of their own death, which is certain unless they take particular measures to avoid it, will spend everything they have in the attempt. To talk of a 'price' of their life in this context is meaningless.

However, most social decisions are not about you or me as individuals, but about the allocation of resources within society as a whole. Cost-benefit analysts are not attempting to discover the value an individual would place on his own life, or indeed that of any other individual; rather, it is the amount the community is prepared to pay to increase its total life-expectancy by one life-unit, by improving the health service or road safety, for example. Whenever society decides to spend more (or less) on lifeboats, mountain rescues, road safety, casualty wards in hospitals or whatever, it is implicitly deciding that it will spend so much money and thereby save, statistically, so many lives. Perhaps it would be valuable to spend the money in order to save the most lives. But whose lives? Is any expenditure on lifeboats justifiable in a world where millions are starving? The problem is that, whilst we are prepared to place limits on the amount spent to reduce death statistics, the moment the people at risk become identifiable individuals, these limits disappear. The miner trapped down the pit, or the sailor lost at sea, have names and personalities which we can identify with, and their predicament arouses our sympathy, and our own fears of death, in a way in which impersonal statistics can never do. Indeed, one of the ways in which we try to understand and come to terms with our own death is to find symbols which absorb it and which, though they do not explain it in any rational sense, nevertheless provide comfort and reassurance. The trapped miner provides such a symbol; statistics do not. When money is spent to rescue him the benefits include not only saving his life, but relieving the anxieties felt by the thousands, and possibly millions, of people who, through the mass media, identify with him. This benefit must be included in any reliable cost-benefit analysis; how to do it in practice, however, is another matter.

This conflict between the needs of the individual and the needs of society runs very deep. Spending money on digging out trapped miners saves less lives than famine relief, but brings greater relief to those who are not directly at risk but are nevertheless emotionally involved. Who is to say that this is wrong; that ordinary people should learn to live with their fears and the Government should get on with the business of keeping the statistics as favourable as possible? Consider another example. Is any expenditure on renal dialysis or geriatric care justifiable, considering the benefits which might accrue to society as a whole from an increased expenditure on the treatment of hernias, piles etc? Indeed, 'should one actually condemn indiscriminate aid in the form of medical, public health, and sanitation services, on the grounds that in spite of their striking immediate results they may in the long run defer the time at which the population can be raised to a stable and adequate standard of living?' (Vickrey, 1953, p.56). The value that society places on human life depends crucially on its willingness to sacrifice the individual for the good of society, a willingness which shows considerable fluctuations with differing circumstances. War time, for example, is very different in this respect from peace time.

Whichever way one turns, the value of human life remains an elusive concept. Indeed, because we cherish other human beings, because human life is so precious, we are very unwilling to place a value on it. Unfortunately, under certain circumstances it is necessary to do so; in such cases, it is important to stress that the valuation used is specific to that purpose alone and cannot be transferred to other circumstances. This will not improve the values used in particular cases, but will guard against the unthinking application of those values to circumstances where their use may be positively harmful.

So far, the discussion has been restricted to the difficulties raised by attempts to produce a relative ordering of values. But the aim of cost-benefit analysis is not only to be able to say which values are better than others, but to quantify those values and the difference between them. Not only does this raise severe technical problems, but also the dangers of presenting the results of the analysis with misleading numerical precision, even though those numbers may be meaningless.

The reality of these dangers is best shown by considering a simple example, taken in part from an article by Martin Gardner, based on a paradox devised by C. R. Blyth (Gardner, 1976). Suppose that there are various drugs which a doctor can use to treat a particular illness. Drug A always produces approximately the same effect. Drug B varies in effectiveness: in 49% of cases it is clearly better than A, but in 51% it is clearly worse and the doctor has no means of knowing in advance which effect it will produce. Obviously, a doctor who wished to maximize his patient's chance of recovery would always choose A, since it is better than B in the majority of cases. Now suppose that a new drug, C, is introduced, which gives the best result in 22% of cases, better results than A but not as good as the better results of B in a further 22% of cases, and worse results than A, though not as bad as the worse results of B, in the remaining 56%. Comparing A with C shows that A is better than C in 56% of cases. Which of the three drugs, A, B or C, should a doctor choose to maximize a patient's chance of recovery? Common sense would say that it is obvious: A is better than B and A is better than C, so A must be the choice. However, a simple calculation shows that A has a 29% chance of giving the best results, C has a 33% chance, whilst B has a 38% chance. In other words, according to this calculation, B is now the best choice! Drug C need never be used; its mere existence reverses the order of preference of A and B. If it is unavailable, A is the best choice; if it is available, then B is the best choice.

Faced with this apparent paradox, the confused doctor, unable to sort out his value judgments and make a rational decision, may turn to the cost-benefit analyst for guidance. The latter will quantify the problem, and may try to rate the effectiveness of these drugs on a scale from one (least effective) to six (most effective). Drug A he may assign the value of three, B either one or five, and C either two, four or six in the proportions given above. The problem is now simple: calculate which drug gives the maximum aggregate score if used in a large number of cases. On this basis, the most effective drug turns out to be C! One begins to sympathize with the befuddled doctor.

The trouble is that the results of the cost-benefit analysis are wholly dependent on the scale which one uses to measure effectiveness. If a different scale had been chosen (and in real problems the choice of scale is rarely obvious but must depend on a number of assumptions and value judgments) then a different result would have been obtained, and either drug A or drug B might have emerged as the most effective. Cost-benefit analysis does not avoid value judgments in such cases; it is crucially dependent upon them. Unfortunately, because its results are presented numerically, it seems to avoid value judgments, and therein lies the danger.

Cost-benefit analysis is an attempt, by analogy with traditional scientific methods, to produce tools which can measure values. So far, it has not been very successful. Inevitably, one asks the questions which were raised in the previous chapter. Are the present failings of cost-benefit analysis merely due to inexperience? Will it be possible in the future to refine this and other economic techniques so that the numbers they provide are meaningful, in the sense that they accurately reflect the underlying ethical values, are reproducible, and can be subjected to error-analysis so that the degree of certainty which can be attached to them can also be quantified? Or are the present difficulties symptoms of much deeper problems? Are the traditional techniques of science simply inapplicable to human problems; are such concepts as the value of human life, even in the statistical sense defined above, inherently unquantifiable? The only way to answer these questions is to keep trying. Cost-benefit analysis is already a useful tool for analysing social problems, and one which is capable of further refinement. But, whilst it may eventually produce more reliable analyses of value judgments, it may also demonstrate that the quantification of certain values must always be arbitrary, and based on political decisions. Until its limitations and the extent of its applicability can be clearly defined, until reliable methods can be developed for assessing the adequacy of its results, it must be used with considerable care, and scientists have both the duty to refine it further, and also a duty to stress its limitations. If they do not do this, then there is a real risk that a wholly unjustified importance will be attached to the results of cost-benefit analysis, elevating it to a folk-science with all the dangers that that implies. Cost-benefit analysis can never obviate the need to make social decisions, and to suggest that it can is dangerous nonsense. But neither is it the worthless tool that many of its critics would like us to believe. Its real function lies somewhere between these two extremes, but where it lies, what value we can place on cost-benefit analysis itself, is still obscure.

References

Cooper, M. H. and Culyer, A. J. (eds), *Health Economics,* Penguin Books (1973).

Ebling, F. J. (ed.), *Biology and Ethics*, Academic Press (1969).

Friedlander, M. W., *The Conduct of Science*, Prentice-Hall (1972).

Gardner, M., 'On the Fabric of Inductive Logic and some Probability Paradoxes', *Scientific American*, 234, no. 3 (1976), pp.119–124.

Henderson, M., 'The Value of Human Life. Cost-Benefit Considerations in Traffic Safety', *Search*, 6 (1975), pp.19–23.

Klarman, H. E., Francis, J. O'S. and Rosenthal, G., 'Efficient Treatment of Patients with Kidney Failure' (1968), reprinted in Cooper and Culyer (1973).

Layard, R. (ed.), *Cost-Benefit Analysis*, Penguin Books (1972).

Miller, H., 'Cost Effectiveness and Problems of Medical Ethics', unpublished.

Mishan, E. J., 'The Value of Life' (1971), reprinted in Layard (1972).

Phelps, E. S. (ed.), *Economic Justice*, Penguin Books (1973).

Vickrey, W. S., 'An Exchange of Questions between Economics and Philosophy' (1953), reprinted in Phelps (1973).

Wolfe, J. N. (ed.), *Cost Benefit and Cost Effectiveness*, Allen & Unwin (1973).

14 Scientists and Ethics

Despite the diversity of topics covered, each of the previous chapters has, at heart, the same theme. Science, whether it is concerned with the esoteric world of theoretical physics or the life and death issues of medicine and world food, is a social activity which must be assessed within the common morality uniting society. All the stresses and strains which underlie that common morality, all the deep divisions and differences of opinion which people have as to what is right and proper and what is not, do not merely impinge upon science. They lie at the roots of the relationship between science and society. Each chapter gives an example of an area within science, or of its applications to social problems, where these stresses and strains rise to the surface and face the scientist, or society, with a difficult ethical choice.

Some of the reasons why these tensions arise can be seen by considering what is meant by 'the common morality uniting society'. Each individual has his own set of values, expressing the importance which he attaches to different aspects of the human condition – life, health, love, liberty, equality or knowledge. These values go very deep; they are a basic response to being alive. Some may be inherent in the human species, others may arise from religious or humanitarian beliefs, or philosophical conviction. Some of these aspects will be valued more highly than others, varying from person to person or from situation to situation. But, wherever they come from and however they change, they form the core of morality. Moreover, despite individual variations, there is normally sufficient overlap that one can speak of a common morality within a society: an agreed set of values and value judgments on which the harmonious workings of that society can be based. Indeed, some writers hold that, in each generation, the common morality precedes the individual; only when he has matured within it can he develop his own responsible variants from it. Inevitably, most individuals will find that there are points where his own personal morality differs from the common morality, and these differences can produce tensions. In extreme cases the result is saintliness or depravity, the zealous reformer or the persistent criminal. More usually, these tensions simply produce frustrations within the individual, a feeling that social standards are under attack, an impatience with the behaviour of others.

Hence, whilst a common morality is essential to a stable and humane society, that common morality is continuously exposed to the tensions within a plural society, and is subject to changing circumstances; and there is always a risk that change will be for the worse rather than for the better. For this reason, morality is not only about values, but also about principles. We do not merely say that we place a high value on human life, we express that valuation in a principle: the sanctity of human life. Principles of this type are essential to good order and behaviour. They provide, in a neat shorthand, the proper basis for our conduct towards other individuals, and towards society as a whole, by establishing expectations of behaviour on which in general all can rely. Without these reliable expectations, the fabric of society is in danger of collapse.

Once a principle has been established, then its application to particular situations gives rise to social aims and objectives. To heal the sick, for example, or feed the hungry, are social objectives which arise from the high value we place on human life and health, as expressed in the principles of the sanctity of life and respect for human personality. But, as many of the previous chapters show, these social aims are diverse: they cannot always be satisfied simultaneously, but may come into conflict with each other. How does one choose, when faced with a new-born child suffering from acute spina bifida, between the preservation of life and the prevention of suffering? How does one balance short-term with long-term benefits in issues such as conservation or space research? Over-emphasizing long-term benefits may condemn our contemporaries to a life of misery; ignoring them may destroy the happiness of our descendants.

Such problems arise partly because different people have different sets of values. Judgments of this type depend upon what sort of person one is, whether one is timid or aggressive, shy or friendly, clever or dim-witted, mystical or down-to-earth, and also on the source of one's morality, whether it is an established faith like Christianity, Islam or humanism, or simply the common morality, tempered by individual conscience. This variety of character implies not only a different attitude to life, but different assessments of the importance of particular values. To take a simple example, some people may value security more than freedom, and hence be willing to adopt more control through the law than those for whom freedom is the more precious. But the conflict may be inherent in the aspects of life being valued. One may value both the candle and the light it produces, but one cannot have both. If these valuations have been enshrined in principles – the preciousness of candles or the necessity of light – then the conflict can seem even more intense. In this simple example, of course, resolution is relatively easy because the principles, as worded, are so shallow as to be almost empty. In such cases, one appeals to deeper principles, such as those of utility or avoidance of waste. But such principles can only be applied because they contain within them an implicit reference to the choice which is necessary. They provide guidance on that choice, but they do not remove it. Even to do nothing is a choice.

Conflict between values, and between different people's sets of values, then, is unavoidable. But the decisions required need not be examined afresh each time such conflict demands it. Experience in handling principles leads to codes of practice: sets of rules, regulations and conventions by which those principles can be interpreted in particular cases. Examples of such codes will be given in a moment. The process of decision is also simplified if a person undertakes an absolute commitment to a particular principle or aim, arising from the very high value which he places on a particular aspect of life. Pacifism is an obvious example. The commitment of many groups to the abolition of experiments with animals is another. Such commitments may be very deep, arising from religious or humanitarian convictions; when they are absolute, then the process of decision is short-circuited and one simply says 'Here I stand, whatever the consequences.' This does not remove the individual's ethical problems, it merely circumscribes them differently. Indeed, it may even increase them. The non-pacifist who, having considered the ethical issues, decides to join up has many of his future decisions made for him; the pacifist, unless he is sent to prison, or is required to offer some equivalent but alternative service to society in place of military service, does not.

However, problems will remain wherever the rules do not provide an obvious answer, and the degree of commitment to both of the conflicting values is similar. In such cases, there is no alternative but to work through the problem, both to try to find deeper principles which one might appeal to, and to consider in depth the consequences of making particular decisions, and how those decisions might affect the interests of everyone involved.

How do these general remarks apply to science? Clearly, the scientist will place a very high value on knowledge since science, in its purest sense, is the systematic search for knowledge. This high valuation will inevitably lead to a position of commitment to knowledge and truth (the extent to which that commitment is absolute will be explored in a moment), and any code of practice which one might suggest for scientists will reflect that commitment. Such codes provide a convenient starting point for a discussion of scientists and ethics.

Bentley Glass has suggested four commandments which should govern the behaviour of scientists: to cherish complete truthfulness; to avoid self-aggrandisement at the expense of one's fellow scientists; to defend the freedom of scientific inquiry and opinion; and fully to communicate one's findings through primary publication, synthesis and instruction. Arising out of these, he suggests three major social and ethical responsibilities: proclamation of benefits; warning of risks; and discussion of quandaries (Glass, 1966, pp.97, 98). Such a code is clearly based on an ethics of knowledge. The scientist must defend his freedom to discover knowledge, and has a duty to communicate such knowledge as he discovers, truthfully, and to interpret that knowledge and its consequences to his fellow man. Almost the whole of Glass's code is based on the notion of the preciousness of knowledge, and the concept of the scientist as the agent by which the community as a whole discovers that knowledge.

At the risk of repetition, one can trace through many of the previous chapters the same thread which runs through Glass's remarks. Eysenck's dictum, that the scientist has a duty to publish the results of his research, for example, is clearly in the same mould. So is much of the 1965 AAAS report, stressing as it does the 'integrity of science' or 'the free pursuit of knowledge'. Moreover, the point was made at the start of chapter 10 that many scientists believe that they must tell not only the truth and the whole truth, but they must tell nothing but the truth. Scientists, according to the Institute of Food Science and Technology, quoted in chapter 5, should avoid 'making apparently authoritative pronouncements outside the area of their particular scientific expertise'.

This kind of formulation has both its strengths and its weaknesses. Its strength is that it recognizes the high value which is placed on knowledge and translates this into a set of simple rules which should govern a scientist's behaviour. Knowledge is a precious commodity, and scientists, as the bringers of knowledge, have special duties and responsibilities which these principles help to clarify. Where they are weak is that they concentrate almost exclusively on one particular value. The purposes behind them are to ensure scientific excellence, to maximize the discovery of knowledge, and to ensure that such knowledge becomes the property of the whole society. They are criteria for moral goodness only in so far as knowledge is the supreme value for man. The many other ethical principles which must affect the pursuit, communication and application of knowledge are not mentioned; yet, as the preceding chapters show, these principles are of great importance. Glass gives no guidance to a scientist who is engaged in defence research, or who wishes to experiment on live animals, or is trying to choose a research project for his graduate students. No mention is made of the ethics of waste, which must affect a scientist's freedom to carry out expensive research. He speaks of 'proclamation of benefits' and 'discussion of quandaries', but fails to distinguish between disinterested informing and committed advocacy.

The point of this argument is not to suggest that the code which Bentley Glass has advocated is wrong, but that it is incomplete. If one considers the pursuit of knowledge in isolation, then one will inevitably arrive at a code very similar to that advocated by Glass. But the practical work of science cannot be considered in isolation from other social values, and these values may conflict with the values and principles inherent in science. The problem then depends crucially upon the relative value which one places on knowledge, as compared with other social values.

Similar observations can be made about other versions of the scientist's code. Cournand and Meyer define five 'norms' for scientists: objectivity, honesty, tolerance, doubt of certitude, and selflessness (Cournand and Meyer, 1976, p.80). These, they rightly argue, are those moral qualities which will ensure scientific excellence, but they also point out that this 'code of the scientist does not contain any implicit or explicit prescription concerning the ways in which scientists should conduct themselves with respect to the application of scientific knowledge to practical affairs', a lack which they try to overcome by suggesting that 'humane generosity' should become a part of this code. But, even here, there is a failure to come to grips with the central problem: that science is a social activity which must take account of other social values; and that a scientist is also a teacher, a therapist, an adviser, an employee and a citizen, not a disembodied concept but a person of flesh and blood who cannot, as Max Walters put it, separate the scientist from the human being.

Within the practice of science, the norms of Cournand and Meyer work quite well. Whenever they are breached, the result is detrimental to science. Most breaches, however, would be morally wrong by any code of conduct. Falsification of evidence, or the deliberate delaying of a paper by a referee in order to gain priority for his own or a colleague's work, are clearly unethical, and against the best interest of science and good behaviour in general. In other cases, however, the element of immorality is less clearly defined. The intense competitiveness which exists in certain areas of science, particularly in the rush for priority of publication, clearly violates the norm of selflessness, yet such competitiveness is acceptable in other walks of life, such as sport. Patenting of discoveries by scientists raises similar difficulties. So can consultancy work, where a scientist undertakes to advise a company, or a government department, for a fee, even though he may believe another would be more competent to advise on the particular problems involved. Then again, many new and revolutionary theories owe their acceptance by the scientific community as a whole to the passionate advocacy of a small number of individuals. For such advocacy to be successful, a scientist may have temporarily to suspend his 'doubt of certitude'.

Here again, the purpose of these arguments is not to suggest that these norms are inappropriate to the pursuit of knowledge, but to point out that science is more than the search for knowledge; it is a social activity whose practices may be more appropriately covered by somewhat different ethical considerations. The value placed on knowledge is central to science, but not so as to make the value of professional or economic reward irrelevant. The latter considerations are clearly operating when the patenting of discoveries is the matter of concern.

This specialized role of these norms becomes even clearer when one considers the role of the scientist in the market place. Consider only one of the norms, namely 'doubt of certitude', in the light of the previous chapters. 'I don't know' or 'I can only estimate' are statements which many laymen find hard to accept: they want to know the facts, and to be sure of those facts. The level of protein required for an adequate diet, the safety of insecticides or nuclear reactors, the benefits from space research – these are things that people want to be able to quantify, to pin down to a particular value. The scientist who commits himself violates the norm and, if he is subsequently proved wrong, may harm both himself and public respect for science. The scientist who exercises 'doubt of certitude' may, like Jensen, find his hypotheses being taken for facts; or, like the Teesdale defenders, may find himself unable to play a full role in political debate.

From these arguments, it should be clear that codes which are based on an overriding commitment to knowledge may ensure scientific excellence, but they can be perturbed by other principles and have misleading implications whenever other values become important. In some cases, such as the patenting of discoveries, a different set of rules can be invoked, as has already been suggested. But in other cases the problem is more fundamental. For example, when experiments using animals are assessed, then the value which we place upon knowledge may come into direct conflict with our desire not to inflict suffering upon animals. Here the codes which have been outlined above do not give clear guidance, since values other than that placed upon knowledge must be drawn into the reckoning. Sometimes a decision must take the form of a choice between strict adherence to a principle and the calculation of the consequences of particular actions on the interests of those involved.

Take the latter first. Chapter 4 dealt with one of the most difficult ethical decisions which any doctor must face: the decision to relieve suffering by failing to preserve life. The preservation of human life is one of the fundamental aims of medicine, and yet, in that chapter, it was argued that a doctor need not always make every effort to preserve life: that there are extreme cases where medical intervention, in an attempt to arrest the natural processes of dying, may not be justified.

In such cases, the decision is based on the interests involved: those of the child itself, the parents, the hospital staff, other children, and of society as a whole. Concentrating on interests in this way does not remove difficulties. Indeed, it is vital that the principles and aims underlying medicine are not forgotten. The tension which the conflict between the preservation of life and the prevention of suffering provides sensitizes the discussion and prevents easy decisions, taken by ignoring one of the principles. It also prevents treating temporary compromises as final solutions. But a balancing of interests does provide a rational basis by which different cases can be compared and appropriate decisions taken for each.

The importance of this approach can be illuminated by dwelling, for a moment, on the nature of responsibility. Even if one adheres to a particular principle, such as the sanctity of human life, and makes all decisions in accordance with it, this still does not remove responsibility because one has to interpret that principle in the light of the particular circumstances involved. Responsibility means being ready and able to justify one's actions to oneself, to any injured party, and to society as a whole. Being able to justify a decision means being able to give an account of how it was reached which seems reasonable and acceptable. If the decision taken was to stick to a particular principle, then one must be prepared to demonstrate why the interests of other people which have been harmed by the decision were insufficient to invalidate the application of that principle. Being responsible, in short, does not mean only sticking to principles; it also means taking into account the interests of everyone involved, impartially balancing one against another and assessing how those interests would be affected by particular decisions.

But this kind of calculus is only a part of the answer. Not only is it necessary to balance the interests involved, but it is also necessary to examine the relative importance of different values, and the degree of commitment which this implies to different courses of action. Some relative values are straightforward, and widely accepted. Most people, for example, would agree that killing one's neighbour is far more serious than coveting his ox. But other values are more difficult to relate to one another, partly because they change from person to person and from situation to situation, and partly because of the intrinsic difficulties in comparing values.

Both chapter 2 and chapter 11 explore one such area of difficulty which is fundamentally important to scientists. Objective knowledge and human happiness are both of value to man. Science, in both the pure and applied sense of the word, is an activity which can contribute to both. Moreover, in the majority of cases, there is little or no apparent conflict between these two objectives of science. Many scientists would argue that the free and unfettered pursuit of objective knowledge is not only the best but the only way in which science can contribute to human happiness. Indeed, one might argue that this contrast is a mistaken one since objective knowledge is an essential prerequisite to human happiness and, for that reason alone, is the more precious. But warm and secure human relationships are also essential to human happiness, and the point made in those two chapters is that the pursuit of knowledge can, by affecting other aspects of life such as this, decrease happiness. In such cases, one's course of action crucially depends upon which one regards as the more important: knowledge, or human happiness.

This is what Jacques Monod said on this subject in 1967.

The only goal, the sovereign good, is not, we must admit, the happiness of man, even his temporal power or comfort, not the Socratic Know Thyself – it is objective knowledge itself. This is a rigid and constraining ethic which, if it respects man as the supporter of knowledge, nevertheless defines a value superior to man himself

(Monod, 1968, quoted in Cournand and Meyer, 1976. p.92.).

This position, which is not uncommon amongst scientists (and indeed is that of one member of this group), is very near to being an absolute commitment to knowledge. As such, it has profound implications for, say, the publication of scientific results. It implies that the publication of such results should be almost exclusively concerned with the advancement of knowledge; the other interests which may be affected, the individual unhappiness, social unrest and even risks to life and limb which might arise from that publication, should have little or no influence on the decision to publish.

Other people, including some scientists, would more nearly equate the value of human happiness and objective knowledge, or even reverse the order as stated by Monod. For them, the issue is not one of commitment but one of calculus: decisions in particular cases must be based on the kind of analysis of interests and consequences which was discussed in chapters 2 and 11. Neither they nor Monod are necessarily right or wrong; this conflict goes so deep that we must learn to accept it and live with it. Ignoring it, or adopting apparently simply solutions, may in the long run be detrimental to both the well-being of society and the practice of science itself. It is for this reason that any position of commitment, either to knowledge or human happiness, must be undertaken in full knowledge of the consequences of that position. Failure to do this over-simplifies the problem and may, for that reason, be morally reprehensible.

When conflicts over value judgments go as deep as this, then the practical questions of ethical decision become important. Who should take particular decisions? To whom is he accountable? How can he obtain advice and guidance? What role does institutional control, including the law, play in establishing and maintaining standards?

One profession where these practicalities are clearly defined is that of medicine, which relies heavily on a complex set of conventions backed up where necessary by written codes and disciplinary procedures. The rules and regulations governing clinical research are a case in point. In every research project decisions have to be taken by the individual doctor or researcher and the experimental subject involved. But a structure exists to provide advice and guidance, and to exert a degree of control. Many hospitals or regions convene ethical committees, consisting of both doctors and laymen, before whom any proposed research project must go to be discussed and vetted. The professional societies, such as the Royal College of Physicians, lay down relevant guidelines. The Medical Research Council, which provides much of the funding for such research in this country, insists not only that any application made to it for the support of clinical research must have the approval of the local ethical committee, but that the researcher must adhere to the policies laid down in the MRC's own guidelines. And the law of the land provides an ultimate control. This hierarchical structure is designed to give the individual doctor maximum guidance, to spell out his legal and moral responsibilities and to establish a generally agreed policy, whilst at the same time safeguarding the interests of the patient. The structure is elaborate, but that is an inevitable consequence of the complexity, sensitivity and importance of the issues involved.

Similar structures exist to cover other types of research, such as experiments on animals, but they are generally less elaborate and, in some cases, less successful. Guidance is offered by some of the professional societies, often covering specific problems such as consultancy work, or relationships with the media. But accountability is much more diffuse. Science is not a unitary profession: the scientist lacks the kind of clearly defined relationship with society that the solicitor or doctor enjoys, and to talk of establishing a register of scientists, from which he can be struck off if he contravenes his professional code of conduct, is probably impracticable. Accountability within specific disciplines – industrial chemistry, veterinary practice or civil engineering, for example – might be more realistic, but still has its drawbacks. Institutional control, which may be necessary for specific, major issues, such as protecting the well-being of patients or preventing the abuse of live animals, is not always a sufficient safeguard.

Having said this, however, there are two areas where the scientific institutions have an important role to play. One is in the production of codes of practice. Despite their limitations, and the near impossibility of enforcement, codes of scientific conduct play an important part in the maintenance of standards within science, and the adoption of a particular code by a professional society makes adherence to that code more likely. The second is a supporting role. If a scientist wishes to point to the social dangers of some scientific activity then he may, particularly if he works in an industrial environment, find that such action endangers his career. In such cases, he must look to his professional society for support, and the latter should be prepared to give it.

Ultimately, however, whilst guidance to scientists in ethical matters is obviously desirable, every problem is different from every other problem, and adherence to codes will not answer them all. For many of the problems raised in this report, there is no substitute for sensitivity and awareness on the part of individual scientists. They are the ones who make the decisions, they are the ones who must be able to grasp the issues involved.

Summarizing the arguments of this chapter, the position of the scientist can be described in the following way. The scientist has, by his calling, a commitment to knowledge and truth, arising from the high value which he places upon them. This commitment is expressed in his codes of practice which attempt to ensure scientific excellence, and also to emphasize his reponsibilities to openness about knowledge and effective communication. Some of the ethical issues which arise in science are dealt with quite effectively by these codes. Others, however, are not, and this is always the case whenever values other than that placed upon knowledge need to be included in the reckoning. The scientist must be aware that, even in the practice of his craft, he is implicitly making value judgments which may have ethical implications. Moreover, whenever his activities affect the community at large, then he has special responsibilities which he cannot abrogate. He must weigh the probable consequences of his work, even though he cannot be certain that he can foresee them all, for good or ill. In all such cases, whenever the conventions of science become inadequate and other values need to be considered, it is important that decisions are consciously reached, and that due weight is given both to the underlying principles and values involved, and also to the consequences of particular courses of action.

This kind of analysis sheds light on many of these problems. A scientist, or society as a whole, may discover such strong commitment to one particular course of action that other values seem relatively unimportant. In other situations careful consideration of the consequences of particular decisions, and their effect on the interests of those involved, may clarify and resolve an apparent conflict of principles. But some conflicts will emerge as inherently unresolvable, and there is then no alternative except to learn to live with them. In such cases it is important that those who are directly affected by the issue should have the major say in deciding how it is dealt with, though at the same time discussion with others is valuable and important.

The extent to which any of this happens depends, among other things, on the degree of ethical awareness amongst the scientists concerned. Many scientists are unwilling to recognize the ethical dimension in science, and some of the reasons for this have been explored in previous chapters. Much of the trouble lies within a scientific education, where there is considerable scope for modifications which would enrich the potential scientist and give him greater ability to cope with the real and sensitive issues which will inevitably confront him in later life. At present, if past experience is any guide, he is ill-equipped for such problems.

The social responsibility of scientists is a much-used phrase in recent years, though its true significance is not always fully appreciated. Social responsibility does not imply any one particular commitment, but an appreciation that science must conform to a common morality, and that the scientist must be aware of the social implications of what he is doing. Ethical understanding, and the recognition of values, are essential to that social responsibility; without a sensitive awareness of the ethical dimension, without a willingness to recognize that values can conflict and that value judgments lie at the foundations of everything we do, social responsibility becomes an empty phrase.

References

Cournand, A. and Meyer, M., 'The Scientists Code', 14, no.1 (1976), pp.79–96.

Glass, B., *Science and Ethical Values*, Oxford University Press (1966).

Monod, J., 'Leçon inaugurale au Collège de France', 3 November 1967, Collège de France, Paris (1968). See also *Chance and Necessity*, Collins (1972).

Suggestions for Further Reading

Bronowski, J., *Science and Human Values*, Penguin Books (1964).

Conant, J. B., *Scientific Principles and Moral Conduct*, Eddington Memorial Lecture, Cambridge University Press (1967).

Dunstan, G. R., *The Artifice of Ethics*, SCM Press (1974).

Dunstan, G. R., *Duty and Discernment*, SCM Press (1975).

Edel, A., *Ethical Judgement: The Use of Science in Ethics*, Glencoe (1955).

Edel, A., *Science and the Structure of Ethics*, Chicago University Press (1961).

Epling, F. J. and Heath, J. W. (eds.), *The Future of Man*, Academic Press (1972).

Flew, A., *Evolutionary Ethics*, Macmillan (1968).

Glass, B., *Science and Ethical Values*, Oxford University Press (1966).

Harre, R. (ed.), *Problems of Scientific Revolution: Progress and Obstacles to Progress in the Sciences*, Oxford University Press (1975).

Horowitz, I. L. (ed.), *The New Sociology*, Oxford University Press (1964).

Huxley, T. H., *Evolution and Ethics*, Romanes Lecture, Macmillan (1893).

Laszlo, E. and Wilbur, J. B. (eds), *Human Values and Natural Science*, Gordon & Breach (1970).

Mackay, D. M., *The Clockwork Image*, Inter-Varsity Press (1974).

Monod, J., *Chance and Necessity*, Collins (1972).

Popper, K., *The Open Society and its Enemies*, Routledge & Kegan Paul (1945).

Purver, M., *The Royal Society: Concept and Creation*, Routledge & Kegan Paul (1967).

Ramsey, I. T. (ed.), *Biology and Personality*, Blackwell (1965).

Ramsey, I. T. and Porter, R., *Personality and Science*, CIBA Foundation (1971).

Thorpe, W. H., *Biology and the Nature of Man*, Oxford University Press (1962).

Thorpe, W. H., *Science, Man and Morals*, Methuen (1965).
Waddington, C. H., *Science and Ethics*, Allen & Unwin (1942).
Waddington, C. H., *The Scientific Attitude*, Penguin Books (1948).